QR   Margalith, P. Z.
76.7
.M37   Pigment
1992   microbiology.

$82.00

DATE			

BAKER & TAYLOR

Pigment Microbiology

Pigment Microbiology

P. Z. Margalith

Faculty of Food Engineering and Biotechnology,
Technion — Israel Institute of Technology,
Haifa, Israel

 CHAPMAN & HALL
London · Glasgow · New York · Tokyo · Melbourne · Madras

Published by Chapman & Hall, 2–6 Boundary Row, London SE1 8HN

Chapman & Hall, 2–6 Boundary Row, London SE1 8HN, UK

Blackie Academic & Professional, Wester Cleddens Road, Bishopbriggs, Glasgow G64 2NZ, UK

Chapman & Hall, 29 West 35th Street, New York NY10001, USA

Chapman & Hall Japan, Thomson Publishing Japan, Hirakawacho Nemoto Building, 6F, 1-7-11 Hirakawa-cho, Chiyoda-ku, Tokyo 102, Japan

Chapman & Hall Australia, Thomas Nelson Australia, 102 Dodds Street, South Melbourne, Victoria 3205, Australia

Chapman & Hall India, R. Seshadri, 32 Second Main Road, CIT East, Madras 600 035, India

First edition 1992

© 1992 P. Z. Margalith

Typeset in 11/13 Palatino by Blackpool Typesetting Services Ltd, Blackpool
Printed in Great Britain at the University Press, Cambridge

ISBN 0 412 41050 8

A catalogue record for this book is available from the British Library

Library of Congress Cataloging-in-Publication data available

This volume is dedicated to the memory of
the late Professor G. Zimmermann, a fine
teacher and a good friend.

Contents

Any color, so long as it's red
is the color that suits me best;
though I will allow there is much to be said
for yellow and green and the rest.

Eugene Field

Preface

Previous volumes in this series have concentrated on several focal issues of microbiology hitherto not covered specifically in the literature. Most of the material can be found scattered throughout scientific journals or reviews but is not available to the interested reader in an integrated yet readable form. Thus, the contribution of micro-organisms to the formation of flavour in foods, an aspect of great importance to the food technologist, was reviewed in the first volume (*Flavor Microbiology*, Margalith, 1981) to be followed by a treatise on the significance of the steroid molecule in microbial life and its effect on certain physiological processes essential to the pharmacologist and fermentation technologist (*Steroid Microbiology*, Margalith, 1986). In attempting to write a third volume in this series, it soon became clear that pigments, although produced abundantly by microbes, have so far received little attention in the fermentation industry.

Since considerable material concerning this aspect of microbial activity can be found dispersed in many different journals, including a number of excellent reviews, an attempt has been made to gather this information under one cover, so that microbiologists and biotechnologists may have easy access to this colourful and stimulating subject. It is hoped that this overview will contribute to further research in this area, as well as to renewed attempts to utilize this information for eventual technical and commercial exploitation.

The impetus to work on the third volume in this series came to me in quite an unexpected way. Having had the opportunity to visit Mexico during a sabbatical leave of absence, I spent some time touring the fascinating sceneries of the Teotiuacan pyramids, not far from Mexico City proper. As in numerous tourist centres of the world, we were surrounded by a crowd of souvenir dealers attempting to sell their merchandise of statues and picture postcards. Trying to evade this human blockade, our guide directed us towards another dealer, huddling in a corner of the main entrance to one of the pyramid shrines. Our attention was immediately attracted by his handling of a 'cactus (opuntia) leaf'. Cacti are very abundant in this part of the world, so we immediately wanted to find out what he was doing. Coming closer, we could easily distinguish the abundant white aphids that covered the opuntia blade. Scraping off some of these creatures and squeezing them against a sheet of white paper, a golden-yellow liquid became apparent. Becoming absorbed in such an interesting demonstration of how colour was manufactured in ancient times, such as how the Tyrian purple was produced from certain molluscs by the ancient Romans, it also occurred to me that much smaller organisms produce pigments of a variety of colours. I soon found out that no full length, comprehensive work has been published on the subject of microbial pigments. It seems to me very appropriate to use this opportunity to thank this anonymous souvenir dealer for having led me to the central theme of this volume.

Although the topic of this work was easily defined, it became increasingly difficult to decide upon the nature and content of its chapters. Pigments are practically ubiquitous in the biological world animals as well as plants produce or contain a large variety of colouring material. Microbes are well known for their biosynthetic capacity and contribute significantly to this class of chemicals often not encountered in other organisms. Since it would be too lengthy to describe the occurrence, chemistry and biosynthesis of all pigments produced by microbes or thallophyte plants, some limitations had to be delineated. There would be little point in discussing in great detail pigments of microbial origin that also occur abundantly in a variety of plant material.

Beta-carotene produced by certain fungi is of course not different from β-carotene formed in higher plants. However, it would be interesting to examine and discuss certain aspects such as its specific functions as a hormone precursor, its role in structural material, or biotechnological trends. Similarly, bacteriochlorophylls will be discussed with only minor reference to the photosynthetic pigments of higher plants. Occasionally, I shall also review observations made with microbial material that turned out to be of a more general, biological significance.

It is customary to distinguish between two categories of microbial products: primary metabolites, those products, formed in larger or smaller quantities, which are essential for the maintenance and reproduction of microbial cells; and secondary metabolites, or idioliths, which are usually produced in minute amounts under certain environmental conditions, but not others, and for which there is little or no evidence for their physiological significance. It should be stated here that microbial pigments must be considered both as primary and secondary metabolites. Here we encounter another difficulty. Many compounds extracted from the growth medium or microbial biomass are somewhat coloured. For example, many antibiotics in crystalline form, or even in concentrated solutions, may absorb the visible range, and could therefore be classified as pigments. In order not to get entangled in the endless number of such compounds, it was decided to limit the discussion to the more abundant compounds, widely recognized as microbial pigments or having unique functions in microbial life, without pretending to cover exhaustively all the available material.

Of course, pigments in biological material have been discussed by earlier workers, such as in Goodwin's *Chemistry and Biochemistry of Plant Pigments* (1976) or Britton's *Biochemistry of Natural Pigments* (1983).

However, in this volume an attempt is made to concentrate on pigments found and studied in the microbial world, with an emphasis on their taxonomic, physiological and biotechnological uniqueness.

Acknowledgements

It is a pleasure to acknowledge the assistance I have received in the preparation of this manuscript. Thanks are due to the librarians of the University of California, Santa Barbara and of the Department of Food Engineering and Biotechnology of the Technion, Israel Institute of Technology, Haifa. In both libraries I was aided in the retrieval of much of the literature upon which this book is based.

Further, I would like to express my sincerest gratitude to Mrs Selma Cartoon who both typed and edited this volume.

Introduction

It is doubtful whether psychologists will agree upon the priority in the importance of colour over shape. In other words, what do we distinguish first, the shape of an object or its colour? It is also possible that the cognizance of colour is dependent on the nature of the object. In all likelihood we shall not be impressed by the colour of the Eiffel tower in Paris, or the Golden Gate Bridge in San Francisco, but one may perceive the colour of a flower long before analysing its morphological peculiarity.

In microbiology, the colour of a culture is usually of great importance; long before establishing the shape of a colony, whether it be smooth or rough, we observe the occurrence of a pigment in the colony itself or its immediate surrounding. This applies to bacteria, and certainly also to fungi when the pigmentation of the mycelium or colourful sporulation elicits not only an aesthetic impression, but is also of considerable taxonomic importance.

The sensation of colour, as we shall see, can be caused by a number of factors. In microbes this is caused almost exclusively by compounds generally referred to as pigments. This can usually be verified by simple solvent extraction procedures and the isolation and study of the spectral qualities of the pigment.

Since this work is written mainly for microbiologists, it will be appropriate to devote some pages to a brief introduction to the principles of photochemistry and to get acquainted with the basic nomenclature employed in the study of pigments. I shall not go into the historical developments that led to the study of their nature and function. However, it is suitable to point out that there is a very old connection between microbes and pigments. In fact, it was Engelman, who in 1880, first demonstrated oxygen evolution in photosynthesizing organisms by his classical microscopical observation that the eucaryotic alga Spirogyra would attract bacteria, which would migrate towards the portion of the algal surface in the neighbourhood of its large spiral chloroplast only when illuminated!

The desire to understand phenomena concerned with colour formation and appearance is not only of importance to the specialist but also to the non-professional observer. Most mushroom collectors know that many Boletus species, when damaged, show a change of colour of the flesh from yellow to blue. What are the pigments involved, and what mechanism induces this change of colour?

Much as we microbiologists like to emphasize the importance of microbes in biochemical reactions or their unique biosynthetic properties, in the case of pigments we shall be obliged to follow a purely chemical order. The structure of the pigment, and not its producer, will be the organizing principle of this volume.

Finally, I would like to use this opportunity to thank the Technion, Israel Institute of Technology, Haifa, for the time and facilities that have been at my disposal to write this work.

1

Photochemistry

Visible light (380–750 nm) is only a small fraction of the spectrum of light emitted by the sun and penetrating through the atmosphere. However, this range of the spectrum is of fundamental importance, not only in making life on our planet possible (photosynthesis), but also in distinguishing between the different colours (vision) which enlighten our environment.

The sensation of colour is created by the absorption of only part of the visible range, and the reflection of that part which is not absorbed. White light is a continuum of the visible range which may be split (e.g. by passage through a prism) into narrower ranges of wavelengths, yielding a series of coloured beams known to us as red, orange, yellow, green, blue and violet. Not all colour sensations have an equal background. Colour may be perceived as a result of light scattering. Particles smaller than the wavelengths of red or yellow light will reflect more of the short wave components of white light and will create the sensation of blue (Tyndall blue), for example the blue of the skies due to light scattering by dust particles of the atmosphere. Similarly, the blue sensation may be related to the light scattering properties of protein particles in the iris of the eye or the air-filled lamellae of 'blue' feathers in some birds. If light is reflected by larger particles, a 'structural' white may be observed (snow white, milk white). Such structural colours are not produced by pigments. Another structural colour may be observed as a result of 'interference', as seen when light strikes a thin oil film on water. In many insects the wing structure is such that a range of interference colours may be observed when viewed from different angles, due to minute air spaces between the laminae of the wing scales. Here also, no pigments are involved.

Chemical compounds with preferential absorption of some of the wavelengths of light and reflection of the rest of the visible spectrum are known as pigments. These are responsible for most of the natural colours in our environment. However, it is not only the colour sensation that is of interest. Far more important is the effect of light (via pigments) on matter – the photochemical reaction (if any) that may be induced by absorbed light.

Let us first review some of the basic laws which control the effect of light on matter. There is a strong relationship between wavelength and the energy of its quanta:

$$E = hc/\lambda \text{ (h = Planck's constant; } \lambda = \text{wavelength;}$$
$$c = \text{velocity of light)}$$

$$v = c/\lambda \text{ (} v = \text{frequency of radiation)}$$

Hence $E = h.v$.

This shows that the higher the frequency of irradiation, the greater its energy content; the shorter the wavelength, the greater the energy of its quanta. Blue light quanta are therefore more energetic than red light quanta, and the invisible ultra-violet quanta are even more energetic!

A few practical principles are relevant to photochemical reactions:

1. Only radiation which is absorbed may be effective in promoting a photochemical change (Grotthus–Draper law).
2. When intensity and exposure time are constant, the photochemical effect is the same (Bunsen–Roscoe law).
3. The fraction of incident light absorbed by a substance in solution is independent of the initial light intensity, but increases proportionally with increase in the concentration of the absorbing substance (Lambert–Beer law).
4. Photochemical changes may not occur in compounds that do not absorb a certain wavelength but may be induced by

other compounds that do absorb at this wavelength. Hence photochemical energy may be transferred from a sensitizing compound to a sensitive compound (Dhar's law). The best example of photosensitization would be the classical experiment of Raab who showed, towards the end of the 19th century, that protozoans placed in a dilute solution of acridine dyes were killed if placed in diffuse, visible light, but not in darkness.

From the laws of chemical equivalence (Planck, Einstein and Stark) we learn that when one quantum of light is absorbed per atom or molecule, one light-activated atom or molecule is produced (primary reaction). What happens thereafter depends a great deal on the nature of the molecule and its environment. Secondary reactions may occur. The absorbed energy may be emitted at the same, but usually longer, wavelengths (lower energy than the wavelength absorbed!) or can cause photochemical reactions, oxidation, photolysis or even polymerization reactions. Temperature has little effect on the primary reactions but may markedly affect the secondary reactions.

Electrons occupy the outer layers of an atom or molecule. We speak of ground state when electrons move undisturbed in their respective orbitals. However, they may be disturbed when electromagnetic energy is absorbed, which leads to the transition from the ground state to an energetically higher, excited level. This is carried out by a quantum whose energy is exactly equal to the energy difference between the two electronic energy levels. Hence, it can be caused only by a certain wavelength of light, the quanta of which correspond to that defined energy difference.

Most of the electrons are capable of excitation. However, the energy required will depend greatly on the nature of the orbital they occupy. This energy will decrease, i.e., longer wavelengths will be suitable, if double bonds occur. If a series of conjugated double bonds are present, the excitation energy will become even smaller, and a situation may occur in which excitation can be promoted by absorption of visible light. As a result, the compound will appear coloured (part of the light being absorbed, the rest transmitted or reflected). As the length of the conjugated double bond system increases, the wavelength of maximum absorption increases with a corresponding change in the colour absorbed. That part of the molecule responsible for this effect is called the 'chromophore'.

Once light is absorbed, the electrons are promoted to a higher energy level, resulting in excited states that are unstable. Electrons tend to give up this energy to return to the ground state. This can be achieved by passing the energy on to some other molecule, or emitting the energy in the form of a fluorescent radiation. Since this is accompanied by the loss of some energy (vibration energy) fluorescence occurs at longer wavelengths. Similarly, the energy may be passed on to molecules which absorb at longer wavelengths.

In an excited state the pair of electrons in the outer orbital may not change its antiparallel spin (around its axis) as occurs in the ground state. This is called the 'singlet' excited state. In some cases, the spin may become parallel (the so called 'triplet' state). If the singlet state is sufficiently long lived, the singlet–triplet transition may take place, resulting in a state much more stable than the original singlet excited state. Also, triplet states may return to the ground state with the emission of energy, usually at longer wavelengths. This is called 'phosphorescence'. This is usually of much longer duration, a number of seconds in contrast to the very rapid emission of energy by 'fluorescence' (about 10^{-8} s). Because of their long lifetimes, triplet states often serve as starting points for photochemical reactions.

For a more comprehensive treatment of photobiology the reader is referred to the textbook by Clayton (1970).

2

Melanin pigments

The black pigment frequently encountered in microbial systems is usually considered to be melanin or a melanoprotein. Melanin and melanin-like pigmentation have been described in various organisms, which may be dark brown, or black only in certain organelles, or may be totally pigmented. In certain cases such intense pigmentation may be of taxonomic value.

2.1 MELANIN CHEMISTRY

Although of common occurrence, the chemical structure of melanin, its biosynthesis and physiological function, have been studied only during the last few decades. Among the natural products, melanin is outstanding in that much work was done on its biosynthetic pathways prior to the elucidation of its chemical structure. It was tacitly assumed that melanin is a polymeric substance although no reliable information was available on the nature of its monomers.

Contrary to all known biopolymers, melanin cannot be solubilized. Chemical degradation may be performed by alkaline fusion at high temperatures (200–250 °C) or by oxidation with permanganate. These reactions yielded comparatively little information but eventually proved to be of considerable importance. In the absence of crystallinity, no primary, secondary or tertiary structure could be ascertained. The only complex materials which have many common features with melanin are the so-called humic acids.

Melanin precursors are metabolic compounds which may be found in various organisms, plants and animals, as well as in many micro-organisms. Compounds serving as melanin precursors may be formed by specific enzymatic reactions in the earlier part of its biosynthetic pathway. However, enzymatic reactions are probably not involved during the later steps and are definitely not concerned with the polymerization reaction itself. Borquelot and Bertrand were probably the first to show, in 1895, that a black insoluble pigment can be formed when tyrosine was acted upon by an enzyme preparation from the mushroom *Russula nigricans*. Such melanins, derived from tyrosine, were later called eumelanins. Other melanins which are not dark brown or black, but have a reddish appearance and were described in the hair and feathers of certain fowls, were found to contain cystein in addition to the derivatives of tyrosine. These were called phaeomelanins. Other melanins, described mostly in lower organisms, having no tyrosine, but a variety of nitrogen-free precursors, such as polyhydroxynaphthalenes, were named allomelanins. Thus, in determining the empirial chemistry of melanin, much could be learned through the alkaline fusion procedure. Enzymatically prepared melanin is probably not completely identical with the natural eumelanins in terms of the content of carboxyl groups (Ito, 1986). However, if prepared in the presence of certain metal ions (Zn, Cu, Co) it appears to be very closely related to natural melanin (Prota, 1988).

Melanins are often regarded as inert substances. This is not entirely true. They display an impressive range of chemical properties by acting as effective redox polymers, ion exchangers and radical scavengers, and by showing a strong tendency to bind with aromatic and lipophylic compounds (Crippa *et al.* 1989).

2.2 MELANIN BIOSYNTHESIS

The formation of melanin, or melanogenesis, is not a single, universal process. Research on various organisms led to a number of biosynthetic schemes. The best known route is that of the

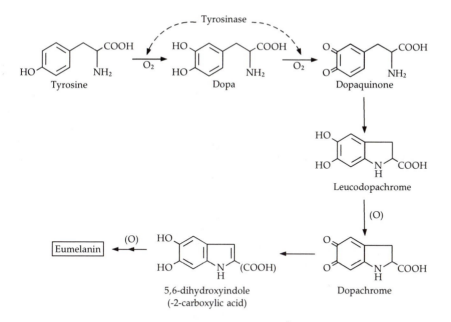

Fig. 2.1 Raper–Mason's pathway of melanin formation from tyrosine by tyrosinase (after Ito, 1986).

Raper–Mason Scheme (Fig. 2.1) whereby tyrosine is oxidized enzymatically to dopa (3,4-dihydroxyphenylalanine) and dopaquinone, which is transformed non-enzymatically to dopachrome (red), 5,6-dihydroxyindole, 5,6-indolequinone (yellow), which undergoes a polymerization process leading to melanochrome (purple) and melanin (black). The empirical formula $(C_8H_3O_2N)_x$ was suggested (Fig. 2.1) (Mason, 1948). This scheme, also known as the 'homopolymeric scheme' was criticized by a number of workers who found that 5,6-dihydroxyindole, prepared enzymatically or by auto-oxidation of tyrosine at pH 8.0, yielded a melanin with an elementary analysis which did not correspond to that of the Raper–Mason data. Also, X-ray diffraction studies revealed no spacing corresponding to repeated identical monomers. According to the 'heteropolymer theory', melanogenesis involves fragmentation of dopa. This was revealed when labelling experiments were conducted. Dopa-carboxyl-C^{14} was transformed enzymatically or by

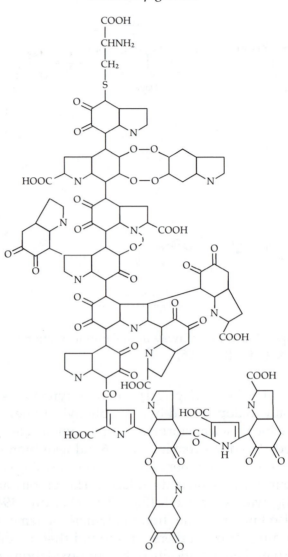

Fig. 2.2 Proposed structures in sepiamelanin (after Nicolaus, 1968).

auto-oxidation into melanin while the CO_2 evolved during the polymerization process was not purely labelled but contained equal amounts of unlabelled CO_2. Such a fragmentation of dopa would not be expected by the Raper–Mason scheme.

According to Nicolaus (1968) who studied the melanin pigment from the cuttlefish (sepia), the polymer was made largely of indole units and some carboxyls, but not all were intact and the monomers were linked to each other in a random fashion by various bonds (Fig. 2.2). Similar conclusions were drawn from data derived from the analysis of melanoma pigments from mice injected with labelled dopa.

The heteropolymer theory has been supported by new evidence which advocates the involvement of free radicals in the polymerization process. Since H_2O_2 is evolved during melanogenesis this seems to be related to the formation of a superoxide anion radical:

$$\text{(Quinol) } HO-\emptyset-OH + O_2 \rightarrow HO-\emptyset-O^- + O_2 + H^+$$

$$O_2^- + O_2^- + 2H^+ \rightarrow H_2O_2 + O_2$$

This is reminiscent of the free radical polymerization reaction in the plastics industry. Such a mechanism would lead to the formation of a heteropolymer. Indeed, inhibitors of free radical polymerization like thiolagents, ascorbic acid and antioxidants, interfere with melanogenesis, while promoters of free radical formation like metals, ultraviolet light and ionizing radiations, enhance pigment production (Blois, 1978). Hence dopa-melanin should be considered as an irregular polymer of various subunits joined by a multiple type of bonding. The subunits comprise 5,6-dihydroxy indole (DHI), 5,6-dihydroxy indole carboxylic acid (DHIC); pyrrole and pyrrole carboxylic acid (Fig. 2.3).

Although the Raper–Mason scheme for the formation of melanin precursors remains essentially valid with all known eumelanins, the biosynthesis of allomelanins (melanins with no or little nitrogen), seems to follow a totally different pathway. Melanins from certain lower plants yield, upon fusion with alkalis, a number of nitrogen-free compounds such as catechol and 1,8-dihydroxynaphthalene. More recent work (Bell *et al.*, 1976; Geis *et al.*, 1984) on melanogenesis of certain imperfect fungi, employing a number of mutants of the melanin pathway, revealed the pentaketide pathway in which

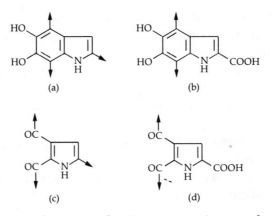

(a)   (b)

(c)   (d)

Fig. 2.3 Suggested structural units present in eumelanin. Arrows indicate possible positions of linkage with each other (after Ito, 1986).
(a) Dihydroxyindole (DHI),
(b) dihydroxy indole carboxylic acid (DHIC),
(c) pyrrole unit,
(d) pyrrole carboxylic acid unit.

acetate is metabolized forming a number of aromatic compounds leading to 1,8-dihydroxynaphthalene, which serves as an immediate precursor for the polymerization of melanin (Fig. 2.4). Allomelanin formation will be treated in greater detail in a later paragraph.

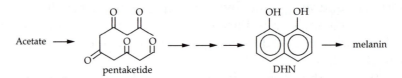

Fig. 2.4 Metabolism of acetate to form 1,8-dihydroxynaphthalene.

2.3 ENZYMES INVOLVED IN MELANOGENESIS

The first step in the melanogenetic pathway, according to the Raper–Mason scheme (Fig. 2.1), is the oxidation of tyrosine to dihydroxyphenylalanine (dopa). This reaction is carried out by the phenolase system, a group of copper enzymes concerned with the oxidation of aromatic compounds which do not display a high substrate specificity. These are mixed function oxygenases; only one atom of oxygen appears in the substrate while the second atom is reduced to water. Thus, tyrosinases will aerobically hydroxylate a phenol in the ortho- position or oxidize a catechol to o-quinone. While one group of phenolases, the cresolases, can catalyse both mono- and diphenols, the other group, the polyphenoloxidases (or catecholases), can oxidize only o-diphenols, having lost completely the monophenolase activity. Both activities are now considered to be performed by one type of enzyme, the catechol oxidases (Mayer, 1987). Another class of phenolase enzymes are the laccases, which may be characterized by their ability to oxidize p-diphenols, in addition to various other diphenols, but will not oxidize tyrosine.

Bacterial phenolases

Bacterial phenolases have been reported in a number of microorganisms, as their cultures form black or dark brown pigments. Melanin was found to be produced by *Aeromonas liquefaciens* and *A. salmonicida*. However, the presence of tyrosinases in eubacteria seems to be a rare event. Aronson and Vickers (1965) described the accumulation of a melanin-like material in cultures of *Bacillus cereus* and *B. thuringiensis* when grown on tyrosine. Cell-free extracts of these bacteria were found to require DPNH for the oxidation of tyrosine. This unusual property has not been confirmed by later work. Pomerantz and Murthy (1974) studied the formation of tyrosinase by *Vibrio tyrosinaticus* grown on tyrosine. Employing cell-free extracts (sonicates) they could show that dopa is required as cosubstrate (the second oxygen acceptor of the mixed function oxygenase) for the oxidation of tyrosine. There was no evidence for the occurrence of tyrosinase in the growth medium. Interestingly, melanin was not formed when *V. tyrosinaticus* was grown on arginine, although the cells produced abundant tyrosinase.

Some *Pseudomonas aeruginosa* strains were shown to produce melanin-like pigments when grown on peptone agar. This would not take place in tyrosine-free media. Since tyrosinase inhibitors (KCN, Na$_2$S) were without effect and dopa could not be identified in culture extracts, there are some doubts with regard to the identity of the pigment. All melanin-forming strains were found to accumulate homogentisic acid (2,5-di-hydroxyphenylacetic acid), while melanin negative strains did not. It was argued that the melanin strains were not tyrosinase positive, but rather mutants defective in the metabolism of homogentisic acid. Under oxidative conditions in the presence of amino acids this was polymerized into a brown aeruginosa melanin (Mann, 1969). A similar brown pigment was shown to be produced by strains of *Serratia marcescens* when cultivated on tyrosine (Fig. 2.5). In the absence of tyrosinase the pigment was not considered to be a true melanin (Trias *et al.*, 1989). Claims to correlate the sporulation of bacilli with the formation of a brown pigment could not be confirmed. Also, here the pigment was not a true melanin (Barnett and Hageman, 1983).

More recently attention has again been directed to the tyrosinase system of *A. salmonicida*, probably due to its economic importance. *A. salmonicida* is considered to be the causative agent of fish furunculosis, especially of salmonids, when cultivated under intensive farming conditions. It produces extracellularly a compound toxic to rainbow trout. When grown aerobically in a tyrosine medium, a highly soluble dark amber to dark brown pigment is formed. Although considered to be a melanin-like pigment, its biosynthesis seems to be somewhat different from the accepted Raper–Mason scheme. Employing labelled tyrosine (2,6-^{3}H-tyrosine) and examining the tritium content of water, it was concluded that the pigment was produced from tyrosine via the formation

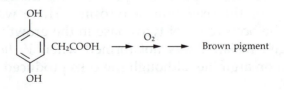

Fig. 2.5 Oxidation of tyrosine to produce a brown aeruginosa melanin.

of 2,3-dihydroxyphenylalanine, 2,4,5-trihydroxyphenylalanine and topaquinone, followed by polymerization to the melanoid pigment (Morgan *et al.*, 1985).

Much more information is available in the literature on the phenolase system of the actinomycetales. With the exception of some work with *Mycobacterium leprae*, most of the publications deal with the Streptomycetes, which have been actively pursued in the search for new antibiotics. With the isolation of virtually millions of Streptomycetes in such screening programmes it has become necessary to find means of identification and classification. Medium pigmentation has been considered an important feature (e.g. Shirling and Gottlieb, 1972). The chromogenicity of many Streptomycetes as manifested by the production of a diffusable dark brown pigment on complex organic media was so significant that it was long considered a key characteristic for the systematics of the Streptomycetes. Mencher and Heim (1962) studied the pigmentation of *Streptomyces lavendulae* when grown on tyrosine agar. The incorporation of labelled tyrosine into the brownish black pigment was considered evidence for the melaninotic nature of the pigment. An assay for tyrosinase activity and melanin formation was later devised by Arai and Mikami (1972). In this assay some culture liquid is added to a substrate solution (pH = 5.9), and after a short incubation the red colour (dopachrome) formed is analysed spectrophotometrically. These workers prefer L-dopa as a substrate, as a number of Streptomycetes failed to show activity on the standard tyrosine agar but were positive with dopa (*S. biverticillatus, S. violaceorectus* and *S. lucensis*). In other species (*S. amakusaensis, S. kentuckensis* and *S. violaceus*) dark brown pigmentation took place on a peptone–yeast extract iron agar, without yielding positive results in the tyrosine or dopa test. This might indicate that not all dark brown pigments of Streptomycetes cultures are indeed melanin pigments. Lerch and Ettlinger (1972) studied the tyrosinase enzyme of *S. glaucescens*, and found that, contrary to other systems, this tyrosinase was a simple copper enzyme and not made of a mixture of isoenzymes.

Attempts to elucidate the cytogenetic aspects of bacterial

tyrosinases have revealed a number of interesting features. Gregory and Huang (1964) working with *S. scabies* claimed that the information for tyrosinase production was carried on a plasmid. On the other hand, Crameri *et al.* (1982) mapped the tyrosinase locus on a chromosome in the case of *S. glaucescens*. Intra- and extracellular tyrosinases were found to be identical, but this enzyme could be induced by methionine.

Bauman and Kocher (1974) showed that two loci were involved in the biosynthesis of melanin in *S. glaucescens*: the *tye* (tyrosine) locus, and the *mel* (melanin) locus, the latter being responsible for the excretion into the medium of the melanin pigment. Thus, on tyrosine medium the *tye+ mel−* genome would not enable melanin excretion. This observation would, of course, explain some of the anomalies and controversial data mentioned earlier.

Cloning of the tyrosinase gene responsible for melanin synthesis in *S. antibioticus* and its expression in *S. lividans* has been reported by Katz *et al.* (1983) employing a plasmid vector. While most activity in *S. antibioticus* was extracellular, the tyrosinase gene in *S. lividans* was expressed only intracellularly. More recently Della Cioppa *et al.* (1990) have shown that a tyrosinase gene from *S. antibioticus* may be expressed in *E. coli* under the control of an inducible bacteriophage T7 promoter. The expression of an additional open reading frame (ORF 438) protein from the *mel* gene locus of *S. antibioticus* was required for high level melanin production in *E. coli*. Recombinant *E. coli* would be an interesting organism for the production of melanin with conventional fermentation techniques.

Melanin-forming genes have also been described in a number of Rhizobium species and are probably plasmid borne. Indeed, extracellular melanin was formed by a number of strains when grown on suitable media. There was some speculation with regard to the necessity of melanin genes for the nodulation process or nitrogen fixation by the symbiont. This could not be verified. The function of melanin in Rhizobia was suggested to be related to the detoxification of polyphenolic compounds which may accumulate in senescing nodules (Cubo *et al.*, 1988;

Hynes *et al.*, 1988). Catechol melanin has recently been claimed to occur in certain bacterial systems. Azotobacter species, under aerobic nitrogen fixing conditions, form dark brown to black pigmented colonies. Shivprasad and Page (1989) have shown that in a Na^+ dependent strain of *A. chroococcum* catechol is both the inducer and the substrate for melanization. Melanin formation seems to be related to high aeration rates. High respiratory rates are known to be necessary for the creation of reducing conditions for protection of the nitrogenase system in aerobic bacterial nitrogen fixation.

Biotechnology

Lincomycin is produced by *S. lincolnensis*, which produces a dark melanin-like pigment in media containing tyrosine. In sulphur limited media this organism was found to accumulate propyl and ethylproline. These two compounds were suggested as intermediates in the biosynthesis of propylhygric and ethylhygric moieties of lincomycin and 4'-depropyl-4'-ethyl lincomycin respectively (Fig. 2.6). L-tyrosine or dopa were found to stimulate propyl- and ethylproline production. This led to the hypothesis that the amino acid portion of lincomycin is derived from the pathway from tyrosine to melanin. Studies with labelled tyrosine supported this contention. It was concluded that seven carbon atoms of tyrosine are incorporated into lincomycin after losing two carbons to yield the propyl-hygric moeity of the antibiotic (Witz *et al.*, 1971).

Sewage pigmentation due to microbial activity may occur as a result of certain chemical compounds. Chlorinated aromatic compounds, widely used as herbicides, may lead to the formation of black coloured sewage due to the degradation by certain micro-organisms. Sewage turns to a dense black colour when exposed to a mixture of 3-chlorobenzoate (3-CBA) and benzoate. Pseudomonas cultures isolated by enrichment with 3-CBA produced the black colour by oxidation and polymerization of accumulated chlorocatechols. Chlorocatechols accumulated in sewage when the pH was below 7.0, but turned into a black melanin-like pigment at high pH values (Haller and Finn, 1979).

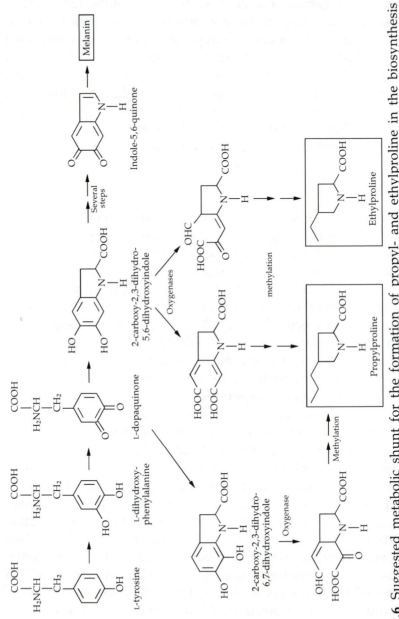

Fig. 2.6 Suggested metabolic shunt for the formation of propyl- and ethylproline in the biosynthesis of lincomycin (after Witz et al., 1971).

Fungal melanins and fungal phenolases

Dark brown or black pigments occur widely in fungi. Except for reactions to wounding, melanogenesis in fungi is restricted to certain developmental stages in special structures like chlamydospores or microsclerotia (Thielaviopsis, Verticillium) or in conidia (*Aspergillus niger*) on hyaline mycelium. In the Dematiaceae, both hyphae and conidia are heavily pigmented (Alternaria, Curvularia, Drechslera). The melanin content can be quite high, up to 10% of the biomass, or up to 20% of the dry cell wall material. Most genera have dark pigments in the sexual spores or their enclosing structures. Melanin appears to be confined to the cell wall region as electron dense grains (30–150 nm). Hence fungal melanins differ from animal melanins which are formed and maintained almost totally as melanosomes and granules **within** the cytoplasm of the specialized cells. Fungal melanins, in contrast, occur either in cell walls or as extracellular polymers formed in the medium around fungal cells. Melanin has not generally been found in fungal cytoplasm, although it has been suggested that melanin precursors may be secreted from the cytoplasm to the cell wall where they are oxidized to melanin (Bell and Wheeler, 1986).

The chemical nature of the fungal melanins was not easily elucidated. Early workers who studied the pigment from the corn smut *Ustilago maydis* spores found catechol, protocatechuic acid and salicylic acid on fusion with alkali. It was therefore suggested that the Ustilago melanin was an irregular polymer made of catechol units of different oxidation stages linked to each other by C—C and C—O—C bonds. Such catechol melanins (allomelanins) contain little or no nitrogen (Piatelli *et al.*, 1965). A similar, but not identical, catechol melanin was described in the pathogenic mold of rice plants, *Cochliobolus miyabeanus* (Ito *et al.*, 1979). However, analysis of the black pigment of *Verticillium dahliae* microsclerotia showed that at least some of the melanins contained indole monomers of the eumelanin type formed from tyrosine or dopa (Ellis and Griffith, 1974). This view was later questioned by other workers who could not detect a stimulation of pigment formation when the fungus was grown on tyrosine. Melanin purified from *V. alboatrum* contained only or less than 1.9% nitrogen versus 6–8% in eumelanins. Since catechol was never isolated from Verticillia this pigment could not be of the type found in *U. maydis*. Work with biochemical mutants proved to be very helpful. Mutant *brm* of *V. dahliae* produced abnormal brown microsclerotia while *alm* mutants formed albino sclerotia. In paired cultures, heterocaryons were obtained that formed the normal black microsclerotia. Since *alm* mutants contained potent phenol oxidizing systems in the walls of cells that normally form melanin, it was concluded that albinism was primarily due to the inability of the fungi to synthesize the normal precursors for melanin. On the other hand, the *brm* mutants synthesized and accumulated a precursor that *alm* mutants could convert to black or dark brown pigments indistinguishable from the natural melanin. The precursor was isolated and identified as (+)scytalone (3,4-dihydro-3,6,8-trihydroxy-1(2H)naphthalenone) (Fig. 2.7). Mutants that accumulate scytalone, but not the wild type, would restore melanin biosynthesis to *alm* cultures when grown in paired cultures (*alm* × *brm*). Since the chemical reaction (Ehrlich reagent) of scytalone mimics that of indole, much of the incongruity regarding *V. dahliae* melanin could be explained. Albino mutants from many black pigment forming fungi (e.g. Curvularia, Alternaria) reacted similarly to scytalone. Neither tyrosinase nor peroxidase oxidized scytalone to melanin. Neither catechol nor dopa are natural intermediates of melanin synthesis in *V. dahliae*. Pigments formed from these substrates by *alm* microsclerotia were atypical in the sequential colour development and other chemical characters. While *brm* mutants could accumulate over 300 mg/l of scytalone, the wild type

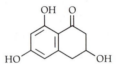

Fig. 2.7 Scytalone.

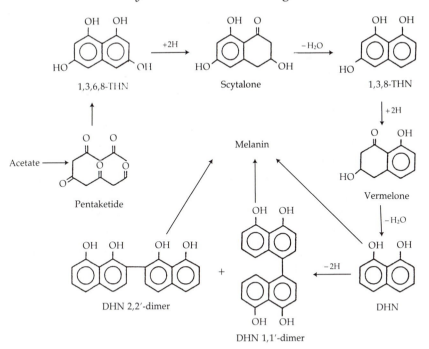

Fig. 2.8 The pentaketide pathway to melanin (adapted from Bell and Wheeler, 1986).
DHN = dihydroxynaphthalene,
THN = tetrahydroxynaphthalene.

produced only minute amounts, indicating the rapid turnover of scytalone in the biosynthetic reactions leading to melanin (Bell *et al.*, 1976).

The discovery of scytalone as an important precursor in the biosynthesis of fungal melanin opened up new avenues in the study of such pigments. Employing a number of artificially induced mutants the pentaketide pathway to melanin could be established (Fig. 2.8). In this scheme, 1,3,6,8-tetrahydroxy-naphthalene gives rise to scytalone, which is dehydrated to 1,3,8-trihydroxynaphthalene yielding vermelone, which is transformed to 1,8-dihydroxynaphthalene, which then undergoes oxidative polymerization to the nitrogen-free melanin.

Contrary to the phenoloxidase systems, the pentaketide pathway is very substrate specific and has been described in a number of fungi: *Thielaviopsis basicola; Pyricularia oryzae* (the rice blast agent); as well as in the human pathogen *Wangiella dermatitidis* (the causative agent of chromoblastomycosis) which synthesizes significant amounts of melanin both in the yeast and, in even higher concentrations, in the sclerotic (parasitic) forms (Geis *et al.*, 1984).

Many Ascomycetes and imperfect fungi appear to produce melanin by the pentaketide pathway. More than 50 species have so far been described (Bell and Wheeler, 1986). However, Basidiomycetes seem to employ a different route in which special substrates are used for the phenolase system that leads to the formation of melanin. Most of the work with mushrooms has been done with *Agaricus bisporus* which produces abundant spores that are heavily melanized. However, no melanization takes place in the mycelium. When grown on horse manure

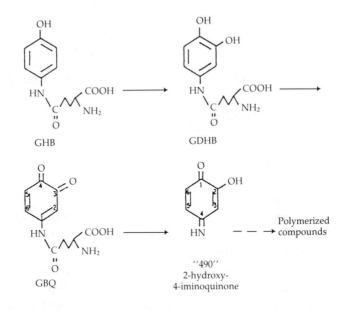

Fig. 2.9 The formation of melanin from GHB in *Agaricus bisporus* (after Boekelheide *et al.*, 1980).
GHB = γ-L-glutaminyl-4-hydroxybenzene,
GDHB = γ-L-glutaminyl-3,4-dihydroxybenzene,
GBQ = γ-L-glutaminyl-3,4-benzoquinone.

compost the fruit body contained high concentrations of γ-L-glutaminyl-4-hydroxybenzene (GHB) which may reach 1–2% of the gill dry matter. Since no tyrosinase activity could be detected in the mycelium while little GHB was found in the basidiospores it was argued that GHB acts as precursor for the synthesis of melanin in the spores (Stussi and Rast, 1981). The following reactions were suggested: mushroom tyrosinase hydroxylizes GHB to γ-L-glutaminyl-3,4-dihydroxybenzene (GDHB), and GDBH is further oxidized to γ-L-glutaminyl-3,4-benzoquinone (GBQ), which then undergoes polymerization to melanin (Boekelheide *et al.*, 1980) (Fig. 2.9). There is good reason to believe that the glutamyl moiety of GHB is not incorporated into the melanin but is removed prior to melanization by a γ-glutamyl-transferase, yielding 4-aminophenol and 4-aminocatechol which are transformed (enzymatically?) to very active oxidized intermediates, such as 2-hydroxyiminoquinone, that polymerize to melanin. The latter compound, also known as '490', which can easily be isolated by simple disruption of the gill tissue, proved to be an excellent sulfhydryl reagent and may contribute to the dormancy of the spores.

GHB is understood to be formed from the shikimic–chorismic acid pathway while the non-aromatic moiety is probably derived from agaritine (Fig. 2.10) which occurs in the fruit body (Stussi and Rast, 1981).

Yeasts are not known to form dark pigments. An exception would be *Cryptococcus neoformans*, a neurotropic pathogen in which the ability to form extracellular polysaccharide capsules and melanin when grown on media containing phenolic compounds (dopa), was found to be necessary in order to demonstrate virulence in mice (Kwon-Chung and Rhodes, 1986). Melanin biosynthesis in this organism was recently shown to follow a modified Raper–Mason scheme for eumelanin (Polachek and Kwon-Chung, 1988). Another case in which

$$HOCH_2-\!\!\!\!\bigcirc\!\!\!\!-NHNHCOCH_2CH_2\overset{\overset{\displaystyle NH_3^+}{|}}{C}HCOO^-$$

Fig. 2.10 Agaritine.

yeast was found to form black pigments has recently been described by Butler *et al.* (1989). From physiological experiments it was concluded that the melanin formed was of the dihydroxynaphthalene (DHN) type. This is a very interesting observation since the black yeast *Phaeococcomyces sp.* is probably a Basidiomycete.

In addition to the catechol, GHB and DHN types of melanins that were described in fungal cells, fungi are also known to produce extracellular (heteogenous) melanins when cultivated on suitable media. These may be formed by two mechanisms: by secretion of phenol oxidases into the environment to oxidize phenolic compounds of various origins (microbial phenols, plant phenols or agrochemicals), and by secretion of phenols into the environment where they are auto-oxidized or oxidized by enzymes released later by the fungus (autolysis).

Mushroom and other fungal tyrosinases form extracellular melanin when tyrosine or protein hydrolysate are available. Such dopa melanins probably have little to do with the melanin of the outer fungal cell wall where most melanin is synthesized. Contrary to tyrosinase, the most common phenoloxidase of fungal cell walls is laccase which does not readily oxidize tyrosine. Further, tyrosinase may be formed in pigmentless (hyaline) mycelia (*Aspergillus nidulans*) while pigmented spores contain laccase (Kurtz and Champe, 1982). Hence, tyrosinase may not be at all related to cell wall pigmentation. Thus, even though tyrosinase was discovered in mushrooms over 90 years ago, there is still no conclusive evidence that dopa melanin is formed in fungal cell walls (Bell and Wheeler, 1986).

2.4 PHYSIOLOGY OF MELANIN FORMATION

While many workers have dealt with the biosynthetic aspects of melanogenesis, comparatively little information is available on the environmental conditions that promote melanin formation in microbial systems. Rowley and Pirt (1972) were the first to examine quantitatively some of the parameters that control melanin production. When *A. nidulans* was grown from a conidial suspension in shake flasks containing a glucose–

nitrate–mineral salts medium, melanin formation took place after the cessation of the exponential growth phase, characteristic of a secondary metabolite. In a chemostat, melanin production was favoured when the growth rate was a relatively small fraction of the maximum growth rate. Growth limitation by the carbon source with a dissolved oxygen tension of 16–135 mm Hg gave the best results. Hence, the restriction of the metabolic rate rather than the concentration of any particular substrate appears to be the factor controlling melanin formation. This would certainly also apply to the melanization of resting organelles as described earlier.

2.5 BIOLOGICAL SIGNIFICANCE

Why do micro-organisms produce melanin? A number of reasons have been offered by various authors. Mostly referring to fungal melanins, it was suggested that hyphae or specific highly pigmented structures may exhibit resistance to a variety of adverse environmental factors during prolonged periods of unfavourable growth conditions or exposure to extreme temperatures. A protective role in desiccation and irradiation has also been suggested. Bloomfield and Alexander (1967) pointed out that such pigmented structures are often responsible for the prolonged persistence of certain organisms in natural habitats. Fungi may be destroyed by bacterial or Actinomycete enzymes that have chitinase, cellulase or β-1,3-glucanase activity. Survival in nature of conidia and sclerotia for prolonged periods may be correlated with the occurrence of melanin and melanin-like compounds in these structures. In fact, the dark hyphae of *Rhizoctonia solani* and *Cladosporium spp.*, both containing melanin, would not serve as a carbon source for soil micro-organisms under experimental conditions. The polyaromatic material of the pigment may act as an inhibitor of cell wall degrading enzymes. When melanized sclerotia or cell walls are attacked by micro-organisms, the non-melanized cells of the sclerotia and the non-melanized portions of the cell walls are generally lysed, whereas the melanized portions of the cells remain intact. It was also suggested that melanin may combine with, or overlay, the lysis susceptible surface components in such a way as to prevent enzymes from combining with their potential substrates. Since polyaromatic materials like lignin and humus are among the natural products most resistant to decomposition in terrestrial and aquatic habitats, this would support the contention that natural melanin pigmentation in microbial systems may confer the ability to survive in environments of biochemically active microflora. This was elegantly demonstrated with melaninless mutants of *A. nidulans* which were highly susceptible to lysis by a glucanase–chitinase mixture, contrary to the high resistance to these enzymes by the pigmented melanin hyphae of the wild type. This observation was further corroborated by *in vitro* experiments in which inhibition of casein hydrolysis by a bacterial protease could be shown in the presence of melanin, the length of the incubation of the protein–melanin mixture prior to the addition of the lytic enzymes negatively affecting the extent of the hydrolytic process (Kuo and Alexander, 1967).

Melanin-containing spores were found to be more resistant to killing by ultraviolet light or solar irradiation. Melanins apparently absorb various types of radiation and dissipate energy primarily by undergoing reversible increases in free radicals. Hence melanized cell walls protect the cell cytoplasm from the damaging effects of free radicals formed by irradiation (Zhdanova *et al.*, 1973). All aerobic organisms form the superoxide radical (O_2^-) and hydrogen peroxide (H_2O_2). Both can do limited damage to cellular targets. However, in the presence of iron catalysts, the interaction between O_2^- and H_2O_2 can result in the formation of the highly reactive hydroxyl radical (OH)$^•$ (the so-called Fenton reaction). Melanins can trap free radicals and thus protect cells against oxidizing conditions (Goodchild *et al.*, 1981).

Melanin formation may also be important in medical mycology. *Cryptococcus neoformans* may produce melanin in appropriate media. In addition to the severe illness due to *C. neoformans* infection in brain tissue, it may also occur in the prostates of AIDS patients, in which it apparently serves as a source of reinfection after successful therapy. Experimental evidence from laboratory infections in mice clearly showed that wild type *C. neoformans* (melanin producer) is much more virulent than albino mutants. The melanin deposited in the cell wall may act as a shield against immunologically active cells (Polak, 1989).

2.6 PHYTOPATHOLOGICAL ASPECTS

We have earlier mentioned a number of phytopathogenic fungi that show dark pigmentation (e.g. *Pyricularia oryzae*). There are many ways in which pathogenic fungi may penetrate into the plant tissue. Hyphae may penetrate through a damaged epidermis or through stomata, or actively penetrate the outer protective layer (cuticle) and reach tissues that supply the nutritional needs of the parasite. This eventually leads to the destruction of the plant host. Active penetration of the epidermal layer seems to be primarily a mechanical process. The hyphae in contact with the outer layer of the host cell wall become firmly attached and usually begin to swell, leading to a depression of the cuticle. Such swollen hyphal tips, or special thin walled structures, are called appressoria. They build up a turgid pressure, enabling a fine hyphal peg to grow directly through the cuticle and the cell wall of the plant (Figs. 2.11, 2.12).

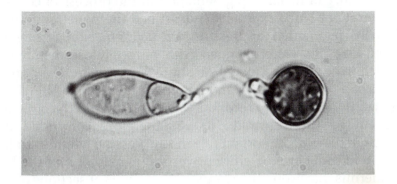

Fig. 2.11 Pigmented appressorium of *Magnaporte grisea* (courtesy of R. Howard, E. I. Du Pont de Nemours and Co.).

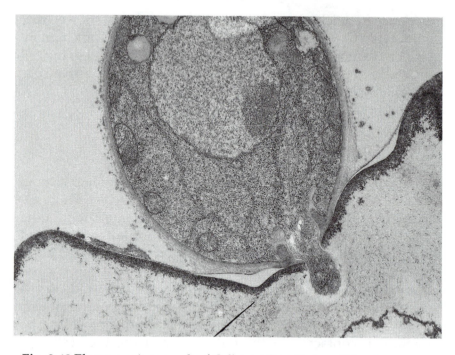

Fig. 2.12 Electron micrograph of *Colletotrichum lindemuthianum* appressoria on detached Bryophyllum leaf epidermis 24 hours after inoculation. Note penetration peg invading epidermal cell. Magnification ×850 (courtesy of H. D. Sisler, University of Maryland, MD, USA).

During the last few years, attention has been directed to the importance of melanin pigmentation in the process of cell wall penetration. Kubo *et al.* (1987) described the penetration from heavily pigmented appressoria of *Colletotrichum lagenarium*. Similar results were obtained with nitrocellulose membranes which were used as a model substrate. When colourless appressoria of albino mutants were examined, there was a lateral germination to form secondary appressoria which rarely penetrated the substrate. It was therefore concluded that melanin plays an important role in suppressing lateral germination and inducing the efficient vertical penetration of the host epidermis. Electron microscopic observations of sections of pigmented appressoria of *C. lagenarium* indicated that the melanized layer extends over the entire appressorium with the exception of the pore which provides an opening for the emergence of the infection peg (Kubo and Furusawa, 1986).

In a recent paper, Howard and Ferrari (1989) pointed out that melanin in the cell walls of appressoria of the rice blast pathogen *Magnaporthe grisea* imparts a differential permeability leading to a high internal solute concentration that creates a high internal hydrostatic pressure, which facilitates the penetration into the host tissue. This observation may be of general importance in the appressorial type of plant infection.

Melanin biosynthesis inhibitors

The epidermal cell walls of aerial plant parts are usually protected by a cuticle layer consisting of mainly cutin polymer (hydroxyfatty acids) embedded in waxes. The cutin–wax association probably constitutes the main obstacle to the entry of pathogens into plant tissue. Many fungi are able to penetrate this layer and infect undamaged plants. During recent years a number of compounds have been developed which show definite antipenetrant activity and thus constitute a class of crop-protecting chemicals. At concentrations not toxic for growth of these fungal pathogens, they interfere with the penetration process. They may be inhibitory to appressoria formation or may interfere with their function in the build up of osmotic forces, affecting the rigidity of the appressoria walls or the adhesive contact with the host. It has to be emphasized that antipenetrants are not specific fungicides and have no effect on postinfectional activity.

So far, the only antipenetrant compounds employed for the control of plant diseases are compounds which interfere specifically with melanin biosythesis (MBI). Among this group, tricyclazole (Fig. 2.13) is being used in the control of rice blast disease caused by *Pyriculuria oryzae*. Tricyclazole blocks melanin biosynthesis at a concentration of 0.1 μg/ml, whereas 20 μg/ml are required for the inhibition of hyphal growth. Activity has been shown to be restricted to those fungi, Ascomycetes and fungi imperfecti, in which the pentaketide pathway for melanin biosynthesis is operative. Tricyclazole

Fig. 2.13 Tricyclazole.

and other MBI compounds (pyroquilon, Fthalide) block the synthesis of the pigment in a number of sites, mainly in the reductive steps (NADPH) between 1,3,6,8-THN and scytalone, and 1,3,8-THN and vermelone (Fig. 2.8).

In the presence of a melanin biosynthesis inhibitor, there was no inhibition of spore germination or appressoria formation. However, the appressoria are not melanized and penetration into the cuticle epidermis does not occur. Melanization seems to provide the rigidity and architecture needed to support and focus the mechanical forces involved in the penetration process (Woloshuk *et al.*, 1983; Sisler 1986). Studying the effect of tricyclazole on the pathogen *Colletotrichum lindemuthianum*, Wolkow *et al.* (1983) observed that the appressoria are often abnormally elongated, suggesting increased flexibility or plasticity of the cell wall in comparison with melanin-containing controls. Hence, antipenetrants lead to a decreased structural rigidity and inability to penetrate the host cuticle, resulting in the efficient protection of the plant against its predator.

2.7 MELANIN IN SOIL

We have earlier mentioned the similarity between melanins and humic acid. Soil humic materials are highly heterogeneous mixtures formed through numerous chemical and biological reactions. Since many dark pigmented fungi and fungal organelles are incorporated in soil, it is logical to assume that these pigments, which display considerable resistance to bio-degradation, may also take part in the build-up of the humus fraction. In addition, extracellular melanins formed by soil-borne fungi by the oxidation of secreted phenols may constitute appreciable amounts of the fungal biomass and may contribute significantly to soil humus. As far as is currently believed, the soil humic acids cannot be considered true phenolic polymers, although they contain these compounds as part of their building blocks, because other widely diverse structures such as polycyclic aromatic hydrocarbons and benzenepolycarboxylic acids are also present. Humic acid is produced by a long process of humification in which melanin and melanin-like compounds are involved. One approach to the evaluation of the humification process is the quantification of the free radicals which would reflect the constitutional changes that occur during the process.

Employing electron spin resonance spectrometry (ESP) Saiz-Jimenez and Shafizadeh (1985) suggested that fungal melanins occur in young or new humus material, in which functional groups may be relatively free of metal ions or interaction with other organic compounds. When incorporated into soils, the melanins and other microbial polymers may play an important role through polymerization–depolymerization reactions with humic compounds and other reactive constituents. However, it may take a very long time before fungal melanins are converted to soil humic acids.

2.8 MELANIN BIOTECHNOLOGY

The formation of melanin may lead to certain economic problems. Several fungi have been described which cause blue or black discolouration of wood. They are not usually wood-destroying fungi, but live predominantly on the protein content of parenchymal cells. *Ceratocystis coerulens* and *Alternaria alternata* have been described as causative agents of wood discolouration (Zink and Fengel, 1988).

Melanoidin pigments also occur in molasses. These pigments are not of microbial origin, but rather are the result of the processing of sugar-containing plant extracts. The enormous quantities of molasses employed in the fermentation industries (yeast, ethanol, etc.) eventually lead to the disposal of these pigments in sewage installations, they being little or not at all degraded during the purification process. Aoshima *et al.* (1985) and Sirianuntapiboon *et al.* (1988) studied the degradability of molasses melanins by various fungi and found that 90% of the pigment could be removed by an undescribed fungus (D90), while among the white rot Basidiomycetes a strain of *Coriolus versicolor* would also show high decolourizing activity. So far, little is known about the end products of such a degradation or the feasibility of its use in the purification of fermentation waste waters.

Much greater attention has been given to the problem of pigmentation and pigment removal (bleaching) in the paper industry. In the Kraft process, lignin, a major constituent of wood, is removed by solubilization and degradation in a solution of $Na_2S/NaOH$ at 170 °C. However, about 10% of the lignin remains in the pulp and imparts the characteristic brown colour of Kraft pulp and papers. This lignin is probably bound to the hemicelluloses and, due to various conjugated structures (including catechols and quinones), appears as a brownish colour. This delignified pulp may now be bleached by a process employing chlorine and its oxide. The enormous amount of effluent produced during bleaching is an increasing environmental concern, and it must be treated before discharge.

Biological bleaching may be a partial answer to the environmental problems created by chemical bleaching and the decolourization of the pulp. White rot fungi like *Phanaerochaete chrysoporium* and other Basidiomycetes are known to degrade lignin. These fungi probably also reduce quinones to their corresponding phenols and thus cause the bleaching of the pulp or decolourize the effluent. A suitable technology for the industrial bleaching process of paper pulp would be of enormous economic value (Erikson and Kirk, 1985).

White rot fungi may also become of practical importance in other areas of water purification. Dyes that are released into the waste water in the textile and dye stuff industries are not readily degraded by conventional water treatments and are therefore considered a serious pollution problem. Azo dyes are the most important colours employed in the textile industry (Fig. 2.14). Under anaerobic conditions the azo linkage may be

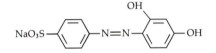

Fig. 2.14 Tropaeolin O.

reduced to form aromatic amines which are colourless. However, these compounds may be toxic and carcinogenic. Recently it was shown that *P. chrysosporium*, especially under lignolytic conditions, may decolourize and efficiently degrade a number of azo dyes (Tropaeolin, Congo Red and Orange II) and may thus become a useful agent in the treatment of effluents in the dye industries (Cripps *et al.*, 1990).

3

The carotenoid pigments

3.1 CAROTENOID CHEMISTRY

More than 500 different molecules of the carotenoid family
have been described. The sophisticated techniques now available, such as HPLC and mass spectrometry not only led to the
discovery of new carotenoid structures but also made it possible to reinvestigate earlier descriptions of carotenoids in a
more reliable manner, often yielding quite different results.

Most of the carotenoids are C_{40} tetraterpenoids made up of
eight isoprene units. The characteristic light absorption of these

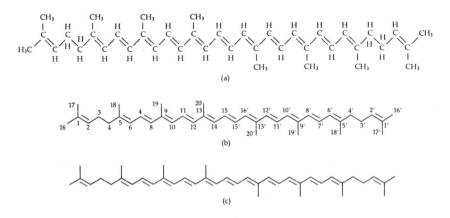

Fig. 3.1 All *trans*-lycopene: (a) structural formula; (b) numbering of
carbons; (c) common formula.

pigments is due to the chain of conjugated double bonds which acts as the chromophore. Carotenoids with less than 40 carbons, as well as longer chain lengths, are also known. The main groups of the most important C_{40} carotenoids may be organized on the basis of their acyclic (Fig. 3.1) or cyclic structure, their *cis* or *trans* configuration (Fig. 3.2) or oxygenation (the xanthophylls: Figs 3.3, 3.4).

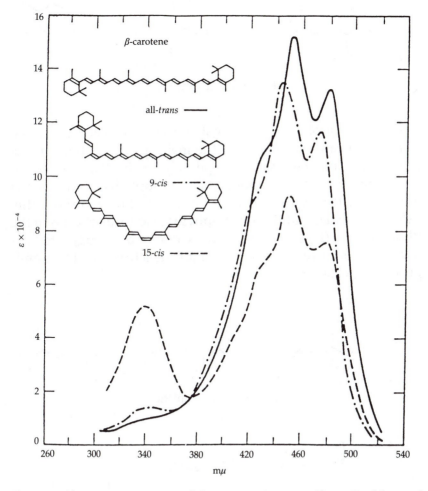

Fig. 3.2 Absorption spectra of β-carotene isomers (from Ruddat and Garber, 1983).

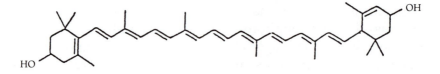

Fig. 3.3 Lutein.

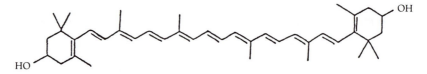

Fig. 3.4 Zeaxanthin.

Some 200 carotenogenic fungal species have been described so far. Beta-carotene is usually the major pigment, but xanthophylls may also be present, especially in the Basidiomycetes and Deuteromycetes. Among the xanthophylls, a number of unusual structures have been found, including aleuriaxanthin (*Aleuria aurantia*) and the xanthophyllic acid torularhodin in the Rhodotorulae (Figs 3.5, 3.6). Some workers have suggested the use of the occurrence of such unique structures for taxonomic purposes (Valadon, 1976).

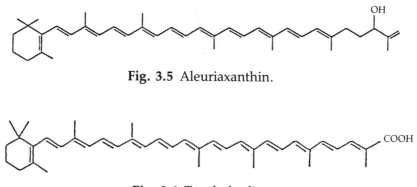

Fig. 3.5 Aleuriaxanthin.

Fig. 3.6 Torularhodin.

I shall not go into the descriptive part of the chemistry of the various carotenoid pigments. This information is well documented and can be obtained from a number of recent textbooks (e.g. Isler, 1971; Goodwin, 1980). I shall, however, briefly discuss their biosynthesis in relation to certain specific processes.

Biosynthesis

Like all terpenoids, carotenoids are biosynthesized from mevalonic acid (MVA) by a head to tail condensation of two isoprene isomers, isopentenyl pyrophosphate (IPP) and dimethylallyl pyrophosphate (DMPP), to form successively geranylpyrophosphate (GPP), farnesylpyrophosphate (FPP) and the C_{20} terpenoid geranylgeranylpyrophosphate (GGPP), which, by tail to tail condensation of two molecules, leads to prephytoenepyrophosphate (PPPP), which is transformed to phytoene (Fig. 3.7). Four successive dehydrogenation steps yield lycopene according to the Porter–Lincoln scheme (Fig. 3.8). Phytoene in fungi occurs mostly in the 15-*cis* configuration. An isomerization step is probably involved leading to the *trans* configuration of phytofluene and the following *trans* carotenoids. Cyclization of the unsaturated acyclic carotenes results in the β-rings. Neurosporene (Fig. 3.8) can give rise to both the acyclic lycopene and the cyclic β-zeacarotene, both yielding γ-carotene. Both pathways may operate simultaneously in many fungi (Bramley and Mackenzie, 1988). A further cyclization step leads from γ- to β-carotene. Oxygenation of the hydrocarbon carotene is the final step that may result in the formation of xanthophylls.

Little is known about the identity of the dehydrogenating and cyclization enzymes, even though some reactions have been demonstrated in cell-free extracts of certain fungi. Activity is usually associated with a particulate fraction.

Carotenes occur predominantly in spherosomes or lipid bodies, but seem to be synthesized in the endoplasmatic reticulum (ER). They accumulate within the lipophilic middle layer of certain sites of the ER. When spherosomes reach a certain size they separate and migrate into the cytoplasm (Wanner *et al.*, 1981).

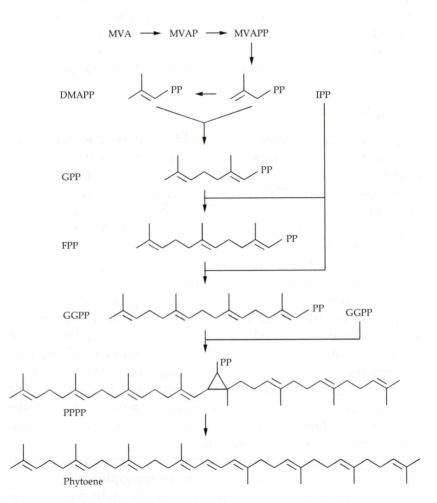

Fig. 3.7 The biosynthetic pathway of phytoene from mevalonic acid (MVA) (from Ruddat and Garber, 1983).
MVAP = mevalonic acid-5-phosphate,
MVAPP = mevalonic acid-5-pyrophosphate,
IPP = isopentenylpyrophosphate,
DMAPP = dimethylallylpyrophosphate,
GPP = geranylpyrophosphate,
FPP = farnesylpyrophosphate,
GGPP = geranylgeranylpyrophosphate,
PPPP = prephytoenepyrophosphate.

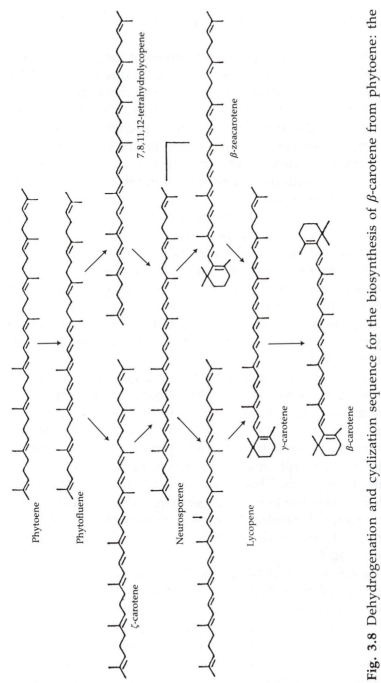

Fig. 3.8 Dehydrogenation and cyclization sequence for the biosynthesis of β-carotene from phytoene: the Porter–Lincoln scheme (modified from Ruddat and Garber, 1983).

3.2 THE FUNCTION OF CAROTENOID PIGMENTS

Of the multitude of pigments formed in the plant kingdom the carotenoids certainly occupy a most unique position. While all photosynthetic plants produce considerable amounts of carotenoids, in microbial systems their distribution is limited to some fungal organisms which have become of great academic interest or even show potential economic importance. Also, in the bacterial world carotenoids may occur in certain groups in connection with photosynthesis or in relation to some unusual ecological features.

The fact that animals are unable to biosynthesize a provitamin they require, while plants do so abundantly, may be related to the fact that green plants are under constant exposure to sunlight, which indicates the importance of these pigments in the process of photosynthesis. The same may be said about the photosynthetic bacteria. This, however, must be modified in those microbes, fungi and bacteria which have no photosynthetic apparatus, yet synthesize certain members of the carotenoid family.

What is the function of carotenoid pigments that makes them a constant component of the photosynthetic apparatus? As mentioned in the following chapter on photosynthetic pigments, two important features should be considered: the contribution to the light harvesting process, and the extension of the light absorption spectrum to wavelengths not absorbed by the chlorophylls. This, however, could not apply to their function in non-photosynthetic systems. The elucidation of their function in heterotrophic organisms may presumably also contribute to the understanding of their role in photosynthetic organisms.

A universal role of cartenoids could be that of a protective action against lethal photo-oxidations. As long ago as 1941 Blum coined the expression 'photodynamic action' to describe the 'sensitization of a biological system to light by a substance which serves as a light absorber for photochemical reactions in which molecular oxygen takes part'. In other words, the photosynthetic pigments are considered as light sensitizers in a reaction that involves O_2 production, which may become lethal to the organism in the absence of a protective mechanism. As we shall see, the necessity for O_2 is not a universal one, considering the photosynthetic organisms that are anoxygenic. Neither are the photosynthetic pigments the only sensitizers that call for photoprotection.

Carotenoid pigments show a protective effect specifically related to the damage caused by irradiation by visible light. From all the available evidence, it does not seem to be concerned with the repair of damage caused by ultraviolet or ionizing radiation. The most common sensitizing compounds are the porphyrins which can also be found in the non-photosynthesizing organisms.

It is very difficult to demonstrate the protective action of the carotenoid pigments in green plants. This is much easier in microbial systems where mutants defective in the biosynthesis of carotenoids can be formed and propagated under conventional light regimes. Working with the non-sulphur purple bacteria *Rhodopseudomonas spheroides*, Stanier's group (Griffiths *et al.*, 1955) produced a mutant strain that differed from the wild type by the absence of carotenoid pigments. This mutant, designated as 'blue–green', accumulates only the colourless phytoene, which in the wild type serves as a substrate to the dehydrogenating enzymes that lead to the coloured carotenoid pigments (Fig. 3.8). When *R. spheroides* is grown under aerobic conditions in light there is an inhibition of pigment formation but growth proceeds normally. Both bacteriochlorophyll and carotenoid synthesis stop. In the case of the blue–green mutant, growth proceeds at a normal rate under either anaerobic conditions in light or under aerobic conditions in the dark. If, however, growth takes place under aerobic conditions in the presence of light, growth ceases and the number of cells decreases sharply, indicating that death has occurred as a result of the photodynamic action. This would not happen when the mutant was grown anaerobically in light or in air in the dark. It is the combined action of light and air that does the killing, in accordance with Blum's postulate of photodynamic action. For obvious reasons, such an experiment could not be performed with a carotenoidless mutant of an oxygenic photo-synthesizing organism. It was further shown that bacterio-chlorophyll was the sensitizing compound, since the blue–green mutant could be grown aerobically in darkness for many generations, leading to a complete loss of the chlorophylls. When such bacteria were exposed to intense light, no detect-able effect on the growth rate could be observed. Hence, in the absence of bacteriochlorophyll, light and air had no effect.

The efficiency of the carotenoid protection against photo-oxidation in photosynthetic bacteria has been studied by a number of workers (Stanier, 1959; Claes, 1960). Employing mutants blocked at various sites of the biosynthetic pathway of bacterial carotenoids as well as diphenylamine (DPA) to block the dehydrogenation step from neurosporene to lycopene (Fig. 3.8), it was concluded that the least unsaturated caro-tenoid which shows a protective effect was neurosporene with only nine conjugated double bonds versus the 11 double bonds of β-carotene and 13 in the spirilloxanthin of *R. rubrum*. With increased desaturation there was better photoprotection (Fig. 3.9).

The protection of non-bacteriochlorophyll-containing hetero-trophic bacteria against lethal photo-oxidation has been the subject of numerous investigations. Many bacteria, especially those exposed to light and air, including aquatic organisms, contain carotenoids, but it is not entirely clear whether such organisms also contain a photosensitizer. However, a photo-dynamic action can be demonstrated if a photosensitizer such as toluidine blue or 8-methoxypsoralen, was added. Working with the carotenogenic *Sarcina lutea*, Mathews (1963) demon-strated a lethal photosensitizing effect, under both aerobic and anaerobic conditions, which did not fulfil the requirements for the photodynamic action as postulated by Blum (1941).

Lethal photosensitization was also demonstrated with the red bacterium, *Serratia marcescens*. The presence of the red pigment prodigiosin (Chapter 8) might be considered as a photosenitizer. However, Muller and Schicht (1965) produced pigmentless mutants which showed little difference in the photosensitivity when compared to the pigmented wild type.

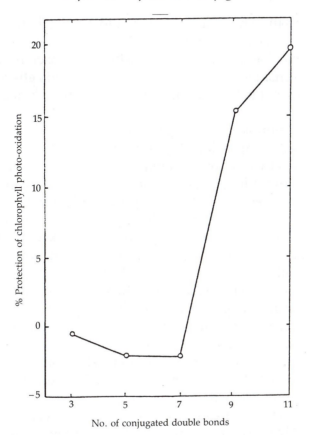

Fig. 3.9 The protective effect of carotenes containing 3–11 double bonds on the photo-oxidative destruction of chlorophyll *a* in petroleum ether after 6 hours of illumination with red light (from Krinsky, 1968).

The question of a sensitizer is still open. A true photodynamic effect could be observed with the carotenogenic *Corynebacterium poinsettiae*. Comparing the sensitivity of the wild type with that of a colourless mutant in the presence of toluidine blue, exposed to 4000 fc of white light for 4 hours in an aerobic environment, the colourless mutant succumbed rapidly, whereas the wild type was unaffected (Kunisawa and Stanier, 1958). This did not happen in an atmosphere of nitrogen. Similar results were obtained with *S. lutea* (Mathews and

Sistrom, 1960). Therefore, what would be the nature of the protective effect of carotenoids in bacteria not exposed to an external photosensitizer? Is it possible that bacteria contain unknown sensitizers which make the carotenoids effective? Or do these pigments have a role not related to the photodynamic action?

Another approach to this problem was that of Wright and Rilling (1963) employing *Mycobacterium marinum*. A photodynamic effect was demonstrated in the absence of an external photosensitizer. Since this organism will synthesize carotenoids only in the presence of light, dark-grown, pigmentless bacteria could be easily compared to the light-grown, pigmented organism. Photokilling of the dark-grown cells took place in very bright light (1400–10 000 fc), provided oxygen was present. Similar results were obtained with the photochromogenic *Myxococcus xanthus* (Burchard and Dworkin, 1966). This organism was found to produce carotenoids when entering the stationary phase of growth, if exposed to light. An attempt was made to isolate the natural photosensitizer. They succeeded in isolating a porphyrin from dark-grown cells (stationary phase) which had spectral and chemical properties similar to that of protoporphyrin IX (Chapter 4). When added to logarithmic phase bacteria (when carotenogenesis was not visible) aerobic photosensitivity could be demonstrated. Moreover, the absence of photokilling during the logarithmic phase of growth could be explained by the minute amounts of the photosensitizer found during this stage.

Comparatively little is known about the nature of the protective effect of carotenoids in photokilling. Since many workers observed that in carotenoidless mutants of *S. lutea*, the killing took place at equal rates at temperatures of 65 °C and 34 °C, a non-enzymic mechanism was suggested. However, when the experiment was conducted with the carotenoid-containing wild type, photosensitivity at 34 °C was less than at 4 °C, which could be due to an inhibition of an enzymic system necessary for maintaining the protective mechanism at the lower temperature (Mathews, 1964). Yet Wright and Rilling (1963), working with carotenoidless strain of Mycobacterium, confirmed the temperature independence of the photodynamic action, but described the protective action of carotenoids to a purely physical phenomenon of shading when the organism was exposed to very bright light.

Macmillan *et al.* (1966) continued this line of research, employing a continuous wave gas laser (at 632.8 nm). In the presence of toluidine blue, photokilling during aeration could be shown with many bacteria (*E. coli, Chromobacterium violaceum, Arthrobacter atrocyanus, Pseudomonas aeruginosa*). This was a true photodynamic effect, since killing would not take place in an atmosphere of nitrogen. They could also show that a pigmentless strain of *S. lutea* was much more sensitive to photokilling than its pigmented counterpart.

Halophilic organisms are very often pigmented and contain various carotenoid compounds. Also, here, a protective action of carotenoids against bright light (12 000 fc) has been shown (Dundas and Larsen, 1962). This could be in accordance with the environmental features of these organisms, which are frequently exposed to direct sunlight.

True photodynamic activity has also been demonstrated with various eucaryotic micro-organisms. Goldstrohm and Lily (1965) showed an increase in sensitivity to light when a non-pigmented strain of the fungus *Sporidiobolus johnsonii* was exposed to sunlight or artificial irradiation. Similar results were obtained with the photochromogenic fungus *Dacryopinax spathularia*. The dark-grown, non-pigmented fungus was exposed to sunlight (2000–70 000 fc) and, after 2 hours, 89% of the organisms were killed, while light-grown cells were fully viable. Also, here the presence of oxygen was a prerequisite for the photodynamic action.

Photokilling was shown with yeast, the pigmentless *Saccharomyces cerevisiae*, as well as the pigmented *Rhodotorula glutinis* (Maxwell *et al.*, 1966) if toluidine blue was used as a sensitizer. However, in contrast to the reports of Dworkin (1958) on bacteria, there was an increased sensitivity of *R. glutinis* with an increase of temperature (3 °C to 40 °C). Since this also happened with white mutants, it could be considered a purely photochemical effect.

Algae also show a photodynamic effect. Bendix and Allen (1962) worked with *Chlorella vulgaris, Chlamydomonas reinhardii* and *Chlamydomonas pyrenoidosa*, and usually found that pigmentless mutants were more light sensitive than the wild types, although some of the strains lost their photosensitivity when grown heterotrophically.

Mechanism of carotenoid protection

Not much is known about the physico-chemical nature of the protective effect of carotenoids during the photodynamic action. In addition to the shading effect mentioned earlier, a number of hypotheses have been put forward to describe the mechanism on a more chemical basis.

In the case of photosynthetic organisms, chlorophyll (Chl) is no doubt the main photosensitizer of the cell. An oxidation reaction has been suggested in which the ground state (g) is sensitized (s) by light and the chlorophyll assumes the triplet state (t) which becomes oxidized and bleached:

$$Chl^g \xrightarrow{h\nu} Chl^s \longrightarrow Chl^t \xrightarrow{O_2} Chl^+O_2 \longrightarrow Chl \text{ (bleached)}.$$

The addition of carotenoid pigments to a chlorophyll solution in petrol ether did protect the chlorophyll from photo-oxidation. Claes and Nakayama (1959) added various carotenoids and showed that the protective effect resulted from the conjugated double bonds. At least nine conjugated double bonds were necessary to protect the chlorophyll from photo-oxidation. Phytoene, phytofluene and ζ-carotene (3,5,7 conjugated double bonds respectively) were ineffective. (Compare this with the results obtained with mutants!) It was argued that, during photoprotection, the carotenoids quench the Chl^t back to the ground state with the dissipation of energy. On the other hand, the oxidation of the carotenoids themselves (5,6-epoxides) may be involved. Krinsky (1964) demonstrated that, during the illumination of *Euglena gracilis*, the antheraxanthin is converted to zeaxanthin (Fig. 3.10) by an antheraxanthin de-epoxidase. Such a carotenoid pair could fulfil the requirements

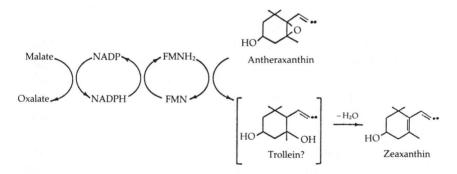

Fig. 3.10 Proposed mechanism of action of antheraxanthin de-epoxidase from *Euglena gracilis* (from Krinsky, 1964).

for the continuous protection against lethal photosensitized oxidation.

Carotenoids have the capacity to quench or inactivate excited states of molecules. This process is best exemplified by the quenching of excited states that are formed in photosensitized reactions. Light converts a sensitizer molecule (S) to an electronically excited form, the singlet species (^{1}S). Excited singlet sensitizers are extremely short-lived and many may be converted to the more stable triplet state (^{3}S). These may react with various molecules and initiate a photochemical reaction. Carotenoid pigments can inactivate triplet sensitizers, regenerating the original sensitizer and forming a triplet state carotenoid (^{3}Car). The triplet carotenoid can return to its ground state with the liberation of small amounts of heat:

$$^3S + Car \rightarrow S + {}^3Car$$

$$^3Car \rightarrow Car + heat.$$

Triplet sensitizers can also react with various biomolecules to generate radical species, then can proceed to initiate radical reactions and cause damage to the cell. Further, the triplet sensitizer can react directly with oxygen to form the excited singlet oxygen ($^1O_2$). This is a very active species leading to the oxidation of various cell components, lipids, enzymes, RNA or DNA, by reacting with unsaturated fatty acids, amino acids or nucleotides. Carotenoids are very effective in inactivating singlet oxygen, the triplet carotenoids thus formed again reverting to the ground state with the liberation of some heat.

Contrary to the quenching of sensitized molecules or singlet oxygen by carotenoids, the quenching of radicals (e.g. in lipid peroxidation) is accompanied by the bleaching of the caro-tenoids, indicating the breakage of the polyene chain of the molecule (Krinsky, 1989).

Protection against non-photo-oxidative damage by caro-tenoids has recently been demonstrated with the pigmented yeast *Rhodotorula mucilaginosa* (Moore *et al.*, 1989). The yeast was cultivated either in liquid culture with duroquinone, a redox-cycling quinone known to generate intracellular super-oxide radicals, or in a hyperoxic atmosphere (80% O_2). Neither of these oxidation challenges affected cell growth unless carotenogenesis was blocked by the addition of diphenyl-amine. Cultures containing diphenylamine were completely inhibited by duroquinone or hyperoxia; only 25% of the cells were found to be viable after 50 hours' incubation. Hence, carotenoids are capable of preventing oxidant-induced cyto-toxicity in *R. mucilaginosa*.

The 'spanner' theory

From the above it may be concluded that, in the absence of a photosensitizer, the protective action of carotenoids against lethal irradiation may not be of great importance. In an attempt to understand the function of carotenoids in heterotrophic bacteria, attention was directed at a different angle.

The cytoplasmic membrane is one of nature's most complex organs. By separating the cytoplasm from its environment, it creates the individuality of the cell. At the same time, it per-forms one of the major functions of living organisms, that of a permeability barrier which enables the selective transport of molecules from the environment into the cell and *vice versa*. This so-called 'semipermeable membrane' regulates the molecular and ionic composition of the intracellular milieu. Membranes are sheet-like structures consisting mainly of

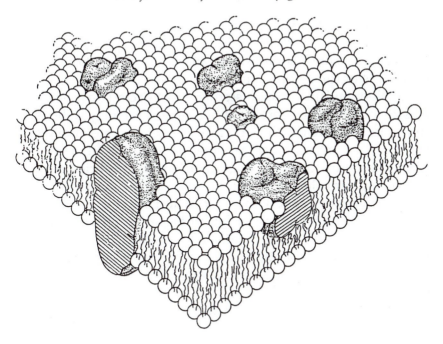

Fig. 3.11 The Singer–Nicolson fluid mosaic model of the cell membrane structure.

lipids and proteins, non-covalently assembled but with a co-operative character. Following the Singer–Nicolson model (Fig. 3.11) the lipids form a bilayer leaflet with the apolar fatty acids pointing towards each other, while those of the more polar terminal moiety (the phospholipids) lie towards the outer (or inner) surface of the membrane. Proteins are embedded at various sites, within or on top of the bilayer.

In the membrane there is a dynamic state in which a lateral diffusion of both lipid and protein takes place. This depends a great deal on the ordered, or rigid, state versus the relatively disordered, or fluid, state of the acid chains of the lipids. The fluidity of the membrane is thus determined by such factors as the nature of the lipid components, the unsaturation of the fatty acids, and their conformation. Important components of the membrane are the sterols which contribute significantly to the rigidity of the membrane. Being comparatively rigid molecules, the sterols will enhance the rigidity of the bilayer. Thus, the fluidity will be maintained by the appropriate ratio between the sterol composition and the unsaturation of the fatty acids of the cytoplasmic membrane.

Not all organisms contain sterols. Most bacteria neither synthesize nor require sterols. Other molecules may substitute sterols in 'reinforcing' the structure of microbial membranes. It has been suggested that carotenoids with about a double length of sterols may serve as rigid 'spanners', bridging and reinforcing both layers of the bilayer leaflet (Ourisson *et al.*, 1979; Rottem and Markowitz, 1979). This could explain some of the earlier observations regarding the sparing effect of sterols on carotenoid biosynthesis in the Mycoplasmas and the replacement of sterols by carotenoids in the sterol-requiring Mycoplasmas (Smith and Henrikson, 1966). Thus, carotenoids may constitute an important structural component of cells unrelated to photochemical reactions.

Carotenoids and antibiotic activity

It is possible that carotenoid pigmentation may also be related to sensitivity to antibiotics. Patients suffering from AIDS have recently been found to disseminate highly virulent Mycobacteria of the *M. avium* complex. Unpigmented variants were found to be highly resistant to therapeutic antibiotics. A selection of unpigmented variants due to antibiotic treatment may be the cause of failure of such chemotherapeutic treatment (Stormer and Falkinham, 1989). So far, no theory has been put forward to explain the sensitivity of pigmented strains to antibiotics. Could this be due to the ordered state of the cytoplasmic membrane as described by the 'spanner' theory?

3.3 MORE ABOUT CAROTENOIDS IN FUNGI

It is a pleasure to observe the growth and development of carotenogenic fungi, especially the Mucorales. The bright yellow of the carotenoids (usually β-carotene) can be observed in the young mycelium and young sporangiophores, as well as in the biomass of a liquid culture grown on suitable media in shake flasks. *Phycomyces blakesleeanus* (Mucoraceae) and *Blakeslea trispora* (Choanopheraceae) are the objects of choice for the demonstration of fungal carotenogenesis in the classroom. Other species, such as *Mucor mucedo, Ustilago violacea, Neurospora crassa* and *Fusarium aquaeductum*, as well as many carotenogenic yeasts (e.g. *Rhodotorula glutinis*), are quite impressive. Since the quantities of carotenoids produced by these fungi are quite considerable, the question may be asked if, in addition to the aforementioned, other functions could be ascribed to fungal carotenes.

Twenty years after Blakeslee (1904) discovered the phenomenon of heterothallism, Burgeff (1924) showed that a diffusable substance is responsible for the initiation of sexual reproduction in the Mucorales. The nature of these compounds, as well as their biosynthetic pathways, has been the subject of thorough investigation during the last two decades (Bu'Lock *et al.*, 1976; Van den Ende and Stegwee, 1971). Also, this is one of the few cases in which some chemical features of heterothallism, that is, the mating of (+) and (−) strains, can be demonstrated.

The spores of *Phycomyces sp.* germinate in a suitable environment and produce a mycelial mat of branched hyphae. Vegetative spores are produced in sporangia borne at the end of very long (up to several centimetres) sporangiophores (the macrophores) or on very short (about 1 mm) microphores. If hyphae of the opposite mating type are present, a complex succession of biochemical and morphological events takes place, leading to the formation of zygospores, the product of sexual interaction (Fig. 3.12). After a long dormancy the zygospores germinate and form a new sporangium containing spores that restart the vegetative cycle.

Before contact between the opposite mating types is established they produce the so-called prohormones ((+) prohormone and (−) prohormone, respectively) which stimulate the growth towards each other, leading to the production of the (pro)gametangia, which eventually fuse and give rise to the zygospores. These prohormones are therefore chemotropic agents involved in the direction of growth of the zygophores (the gametangia-forming hyphae). But the formation of

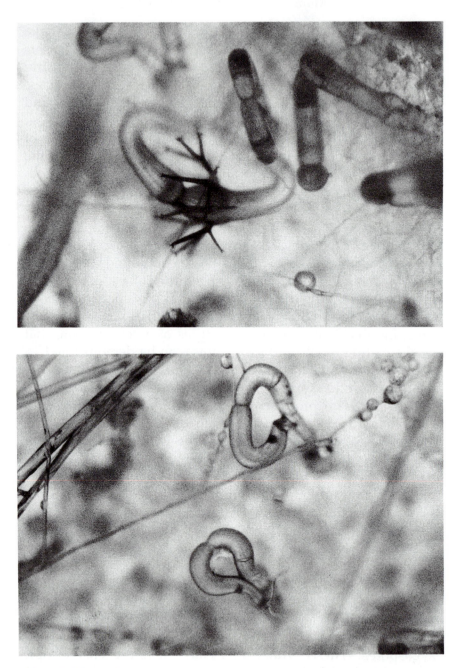

Fig. 3.12 (a), (b) Stages in zygospore formation in *Phycomyces blakesleeanus*.

zygophores and the morphogenetic events that lead to the formation of the gametangia (fusion and zygospore development), are under the control of the hormone itself, produced by the combined action of both mating types (Ross, 1979).

The distinction between the prohormone and the hormone, and their interrelationship, has been the subject of much experimental work and considerable controversy. How can one hormone regulate both mating types? Meticulous chemical analyses in relation to the morphogenetic events could eventually solve this puzzle.

The **plus** mycelium produces the **plus** prohormone which is chemically very similar to the **minus** hormone produced by the **minus** mycelium (Fig. 3.13). Both can act as precursors of trisporic acid, the final hormone, providing some biochemical modification of the corresponding molecules takes place. However, it is the opposite mating type that carries out this event. Thus the **plus** strain forms the **plus** prohormone which is converted by the **minus** strain to trisporic acid by the oxidation of the hydroxyl group at position 4 to a keto group, while the **minus** strain produces the **minus** prohormone, which is also transformed into trisporic acid by the formation of the 1-carboxyl group by the **plus** strain. There are a number of

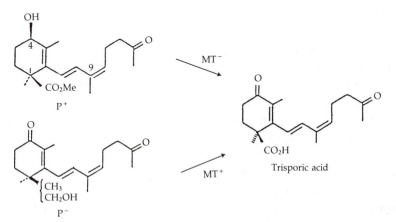

Fig. 3.13 The formation of trisporic acid from the respective prohormones (adapted from Bu'Lock *et al.*, 1976).

trisporic acid molecules, the most important of which is the 9-*cis*-trisporic acid B. Thus, the mutual dependence of each mating type on its counterpart regulates the sexual process.

The brilliant interpretation of a vast amount of experimental data explains the physiological role of the slightly incomplete molecules (the prohormones) as well as the morphogenic function of the common hormone (Bu'Lock *et al.*, 1976; Sutter and Whitaker, 1981a, b).

What has all this to do with carotenoids? Glancing at the structure of trisporic acid, the terpenoid structure is reminiscent of a carotenoid. Indeed trisporic acids are derived from β-carotene as shown by labelling experiments. The transformation of β-carotene (C_{40}) via retinol to trisporic acid (C_{18}) is an essential requirement for the process of sexual reproduction in these fungi. Caroteneless mutants of *Phycomyces blakesleeanus* failed to produce trisporic acids (Sutter, 1975). Colourless, albino mutants should therefore have impaired sexual reproduction.

Mated cultures of (+) and (−) strains produce much higher concentrations of β-carotene than single cultures (Ciegler, 1965). By the addition of trisporic acids to unmated cultures similar results can be obtained. Actually this was demonstrated only with the (−) culture. Since trisporic acid is derived from β-carotene (a positive feedback loop has been implied), the formation of trisporic acid stimulates the biosynthesis of its own precursor (Bu'Lock *et al.*, 1976).

It is possible that the high β-carotene production, as demonstrated under laboratory conditions, is indicative of a biosynthetic potential of β-carotene that may be in excess of the needs for trisporic acid formation for hormonal activity. It is difficult to obtain more information on the quantitative aspects of trisporic acid promotion of the morphogenetic effect, but it seems that trisporic acid is effective in minute amounts. This led some workers to believe that β-carotene may have an additional function. Gooday *et al.* (1973) found a great similarity between the sporopollenin of zygospores (*Mucor mucedo*), a component of the outer part of the cell wall (about 1–3% of dry weight) and that of various pollens of higher plants.

Sporopollenin (C_{90}) is probably synthesized by oxidative polymerization of β-carotene. Hence, carotenoids may contribute significantly to the architecture of the sexual spores of the Mucorales.

An additional role for carotenoids has been suggested for the phototropic response of *P. blakesleeanus* and other fungi. The action spectrum for this response is very similar to the absorption spectrum of carotenoids, which would indicate that these pigments also act as light receptors, both for light-induced carotenogenesis and the phototropic responses. However, a number of findings are not in agreement with this assertion. The induction of the light response by red light (595 nm) as well as by carotenoidless mutants was considered evidence for an alternative, probably flavin, or flavoprotein chromophore. The possibility that two separate photoreceptors may be involved has also been suggested. A response to low-fluence was ascribed to the carotenoid chromophore, while flavins were responsible for the high-fluence effect (Jayaram *et al.*, 1979). This controversy is far from being resolved, although currently most researchers seem to be in favour of flavins and not carotenoids as light receptors. However, direct evidence for the role of flavins in phototropism and carotenogenesis is still lacking (Lipson, 1980; Shropshire 1980).

Interestingly, Lopez-Diaz and Cerdá-Olmedo (1980) distinguish between phototropism and carotenogenesis on the basis of their different threshold values for the photoresponses. The threshold of photocarotenogenesis in *P. blakesleeanus* was about 10 000 times higher than the threshold for phototropism. It seems, therefore, that photocarotenogenesis is geared to much higher light intensities than phototropism. It is likely that the photoprotective role of carotene is needed only at high light intensities, while low threshold effects are limited to the guidance of the sporangiophores for spore dispersion, even in dimly lit environments. It must be further pointed out that phototropic responses are basically differential, the effect depending primarily on the difference between light intensities, which of course cannot be the case with photocarotenogenesis.

3.4 REGULATION OF CAROTENOID BIOSYNTHESIS

Light stimulates the formation of carotenoids. The β-carotene content of Phycomyces was increased over tenfold upon continuous exposure to light in comparison to dark-grown cultures. (Bergman *et al.*, 1973). Light is therefore an important regulatory factor in carotenogenesis before other factors such as mated culture, trisporic acid and nutritional ingredients are considered. Photoinduction has been described in at least 13 fungal species (Rau, 1976) and is probably due to a derepression of the carotenoid operon. However, in a number of cases, a reduction in the accumulation of carotenoids in illuminated cultures has been observed (Sutter, 1969). This is somewhat surprising and may also be explained by photodestruction of the carotenoids.

Light stimulated carotenogenesis may be mimicked by certain chemical compounds. In *Fusarium aquaeductum* and *Cephalosporium diospyros*, which also require light for carotenogenesis, p-hydroxymercuribenzoate and p-chloromercuribenzoate were found to stimulate carotenogenesis in the dark (Parn and Seviour, 1974). It would be of great interest to establish the mechanism by which light stimulates the carotenogenesis. The hypothesis that light inactivates a suppressor of carotene synthesis has not yet been demonstrated experimentally.

Other compounds may enhance light induced carotenogenesis. Since the early days of knowledge of carotenoid production by fungi it has been known that β-ionone has a positive effect (Ciegler, 1965). In spite of the structural similarity to the β-carotene molecule (Fig. 3.14), β-ionone is not incorporated into the product. A regulation in the enzymic activity involved in the conversion of 5-phosphomevalonate to dimethylallylpyrophosphate has been suggested (Rao and Modi, 1977). Since

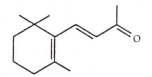

Fig. 3.14 β-ionone.

β-ionone and trisporic acid have a competitive effect, a sensitive site that controls carotenogenesis by gene derepression has been suggested. A 2–3-fold increase in β-carotene production in the presence of certain compounds such as dimethylformamide, succinimide or isonicotynoilhydrazine together with β-ionone has been demonstrated with *Blakeslea trispora* (Ninet *et al.*, 1969). Other compounds have specific inhibitory effects on various sites in the dehydrogenation and cyclization steps of the carotene pathway (Table 3.1). Diphenylamine (DPA) inhibits the dehydrogenation step (phytoene dehydrogenase) leading to the accumulation of the more saturated phytoene intermediate. However, since the total amount of carotenoids increases, one may assume that β-carotene (the end product) regulates its own synthesis (Bramley and Mackenzie, 1988). The fact that certain compounds like nicotine, imidazole or pyridines prevent the cyclization of lycopene makes it possible to employ these inhibitors for the fermentative production of lycopene.

Carotenogenesis is controlled primarily by certain genes that are responsible for the enzymes of the carotenoid pathway. Much of our knowledge of the genetics of carotenogenesis in these fungi is due to the extensive work, mainly with Phycomyces, by the Spanish school (Cerdá-Olmedo, 1985). By preparing a series of mutations with different colours as well as albino mutants, the CarB gene, which determines the dehydrogenation from phytoene to lycopene, could be demonstrated. CarR gene determines lycopene cyclase which converts lycopene to γ- and then β-carotene. Evidently the CarR mutants are red and accumulate large amounts of lycopene. Superproducing mutations have been ascribed to a recessive mutation in gene CarS which leads to the abolishment of the negative feedback control which normally shuts off the pathway when sufficient β-carotene has been produced. This discovery of regulative genes opened up new avenues for strain improvement for carotenoid production. CarS mutants were found to produce up to $6000 \, \mu g/g^{-1}$ dry weight of β-carotene compared to 2000–$2500 \, \mu g/g^{-1}$ dry weight produced by the wild type in the presence of β-ionone.

Table 3.1 Effects of some inhibitors on carotene biosynthesis

Compound	Example	Inhibition of:	Accumulation of:	Reference
Phenylpyridazinone herbicides	Norflurazon (San 9789)	Phytoene synthetase	Phytoene precursors	Sandmann et al., (1980)
Diphenylamine		Dehydrogenation steps	Phytoene	Goodwin (1952)
N-heterocyclic compounds	Nicotine Pyridine Imidazole	Cyclase	Lycopene	Davies (1973)
CPTA herbicide 2-(4-chlorophenylthio) triethylamine		Cyclase with stimulation of carotenogenesis	γ-carotene, lycopene	Coggins et al., (1970)

Employing the ingenious, but tedious, technique of Ootaki (1973) for grafting of young macrophores of *P. blakesleeanus*, stable heterocaryons containing nuclei from two different strains could be obtained. Such heterocaryons showed very high productivity. Heterocaryons produced from mutants of (+) and (−) strains respectively (e.g. S106*M1) yielded up to $25\,000\,\mu g/g^{-1}$ dry weight. This was the subject of a patent granted to Murillo-Aranjo *et al.* (1982).

3.5 FUNGAL XANTHOPHYLLS

A number of oxygenated carotenoids have been described in fungi and yeasts. Fiasson *et al.* (1970) were probably the first to identify the xanthophyll canthaxanthin (β-carotene 4,4'-dione, Fig. 3.15) in the mushroom *Cantharellus cinnabarinus* and related fungi. Xanthophyll compounds such as aleuriaxanthin (1',16'-didehydro-1',2'-dihydroxy β,ψ-caroten-2'-ol, Fig. 3.5) or plectaniaxanthin (3',4'-didehydro-1',2'-dihydroxy-β-carotene-1'-diol, Fig. 3.16) have been described (*Aleuria aurantia*) by Arpin (1968). Generally speaking, the occurrence of xanthophyllic carotenoids in mycelial fungi is not widespread. Interestingly, in yeasts oxygenated carotenoids are quite common. Torularhodin (3',4'-didehydro-β,ψ-caroten-16'-oic acid, Fig. 3.6) can be found in most Rhodotorulae as well as other yeasts.

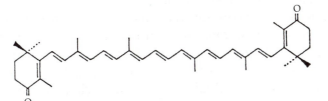

Fig. 3.15 Canthaxanthin.

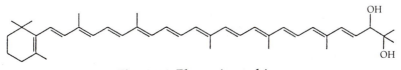

Fig. 3.16 Plectaniaxanthin.

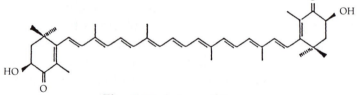

Fig. 3.17 Astaxanthin.

The Basidiomycetous yeast *Phaffia rhodozyma* was found to contain astaxanthin (3,3′-dihydroxy-4,4′-diketo-β-carotene, Fig. 3.17) as its primary carotenoid (Andrews *et al.*, 1976).

Xanthophyllic carotenoids have attracted special attention by workers in the field of pigment biotechnology. There exists a demand for egg yolks with high pigment content. Suitably pigmented egg yolks are needed in the egg industry, both for consumption as well as for bakery goods, noodles, margarine and other foods. The colouring capacity of chicken feeds is of primary importance in the modern poultry industry. Chickens raised in a farmer's yard usually get enough fresh plant material or yellow corn, rich in xanthophylls (cryptoxanthin, Fig. 3.18), zeaxanthin and lutein (Figs 3.3, 3.4) to provide good colouring material for the egg yolk, and often also for the skin or shanks of broilers.

However, in the poultry industry where chickens are raised and fed in batteries, the problem of colouring material raises several difficulties. The use of grains such as milo, wheat or barley, because of economic reasons, does not provide sufficient colour. Fresh plant material cannot be provided in suitable amounts on a large scale poultry farm. Although carotenoids can provide an attractive colour, these pigments are usually converted in the intestinal mucosae of chickens to compounds (vitamin A) which are not deposited in the egg yolk. Therefore, only xanthophylls which are efficiently absorbed can be used as

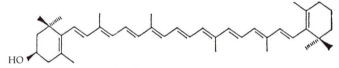

Fig. 3.18 Cryptoxanthin.

colouring material. Dihydroxy- or diketoxanthophylls are better utilized for yolk pigmentation than monohydroxy-, monoketo- or epoxycarotenoids (Braeunlich, 1978).

Attempts to provide xanthophylls of microbial origin in the diet of laying hens have been made by several workers. Schwarz and Margalith (1965) reported on the use of spray-dried *Rhodotorula mucilaginosa* as a feed supplement at 1–2% levels with considerable enhancement of the yolk colour. Higher concentrations resulted in somewhat unnatural red hues. Johnson *et al.* (1980) employed the astaxanthin-producing *Phaffia rhodozyma* for colour enhancement in the diet of laying hens and Japanese quail. By employing broken cells or yeast oil in combination with yellow corn or marigold flower pigments, golden egg yolks were obtained. Also, an orange–pink hue was obtained when high concentrations of the yeast preparation were used.

Currently, synthetic preparations are being employed as colourant agents. β-apo-8'-carotenal (Fig. 3.19) or β-apo-8'-carotenoic acid methyl ester, and canthaxanthin (β-carotene 4,4'-dione, Fig. 3.15) in suitable mixtures with alfalfa meal give good results. However, the current trend against synthetic chemicals in food will undoubtedly encourage new investigations for the use of microbial pigments in the poultry industry (see also bacterial xanthophylls).

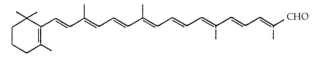

Fig. 3.19 β-apo-8'-carotenal.

3.6 BACTERIAL CAROTENOIDS AND CAROTENOGENESIS

Carotenoids in non-photosynthetic bacteria

Among the natural carotenoids, their occurrence and biosynthesis in bacteria occupy a special position, both from the structural and functional viewpoint, as well as from their evolutionary significance. Bacterial carotenoids do not form a homogenous group. I shall therefore divide them into two categories according to their involvement in chemo-organotrophic and photosynthetic carotenogenesis.

In chemo-organotrophic bacteria, recent research has shown that, in addition to the well known C_{40} carotenoids, C_{30} and C_{50} carotenoids can also be found. The C_{30} compounds, also known as diapocarotenoids (based on the reduced number of carbons at both ends of the molecule) were first described in a strain of *Staphylococcus aureus* (Suzue *et al.*, 1967) and were later identified also in *Streptococcus faecium* and *Pseudomonas rhodos*, as well as in Halococci. A correlation between the G + C content (<59 mol%) and the occurrence of these carotenoids has been suggested (Taylor, 1984). Obviously, questions have been raised regarding the biosynthesis of these unusual carotenoids, as well as their specific function. Are the C_{30} carotenoids derived from their C_{40} counterparts by secondary elimination of two isoprene units, or are they synthesized in an autonomous C_{40} independent pathway? There is good evidence that the C_{15} FPP (and not the C_{20} GGPP) is the immediate precursor and a pathway parallel to the C_{40} carotenoids is operative. Diapophytoene is dehydrogenated to diapophytofluene, which gives rise, through diapolycopene and β-diapocarotene to diaponeurosporene (Fig. 3.20). This is corroborated by the fact that no C_{40} carotenoids have been described in these organisms.

In these bacteria the final product of carotenoid biosynthesis is usually a glycosylated compound. In some cases (e.g. *S. aureus* 209 P) biosynthesis is terminated by oxygenation and the

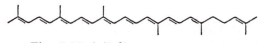

Fig. 3.20 4,4'-diaponeurosporene.

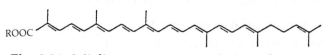

Fig. 3.21 4,4'-diaponeurosporene-4-oic acid ester.

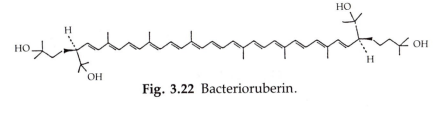

Fig. 3.22 Bacterioruberin.

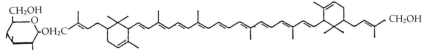

Fig. 3.23 Decaprenoxanthin monoglucoside.

formation of the xanthophyllic glycosidic diaponeurosporene-4-oic acid ester (Fig. 3.21). On the other hand C_{50} carotenoids (e.g. bacterioruberin, Fig. 3.22) in Halobacteria or decapreno-xanthin in Corynebacteria (Fig. 3.23) are synthesized from C_{20} isoprenoid precursors similar to the C_{40} carotenoids with whom they are usually associated (e.g. *Sarcina litoralis, Cellulomonas dehydrogenans, Halobacterium cutirubrum*).

It must be emphasized that C_{30} carotenoids have been found only in chemo-organotrophic and not in photosynthetic bacteria. If mutual exclusion of the C_{30} and C_{40} pathways did evolve, it may be asked why did the C_{30} pathway appear at all? Assuming a similar role for such membrane-associated carotenoids, that is, membrane reinforcement and photoprotection (9–11 double bonds!) or as electron transfer components, it is not clear why those functions could not be addressed by the common C_{40} carotenoids (Taylor, 1984).

Among the chemo-organotrophic bacteria glycosylated carotenoids have been also described in the Myxobacteriales. These are typical soil bacteria with a gliding, creeping movement along suitable surfaces. They are further distinguished by the aggregation of the vegetative rods into characteristic fruiting bodies containing the metabolically dormant myxospores (Reichenbach, 1986). The vegetative mass, and especially the fruiting bodies, of these bacteria are usually heavily pigmented, having yellow, orange, red or purple colours. Carotenoids isolated from these organisms (up to 60 compounds) are of normal

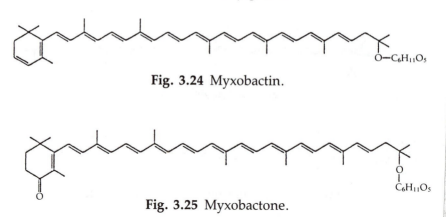

Fig. 3.24 Myxobactin.

Fig. 3.25 Myxobactone.

length, either acyclic or monocyclic (*Myxococcus fulvus*) as well as aromatic (*Nannocystis exedens*). Many contain hydroxyl and keto groups. The hydroxyl groups may be glycoslyated (glucose, sometimes also rhamnose). All myxobacterial glucosides are found to be esterified with a fatty acid (mainly straight chain) at one of the hydroxyl groups of the sugar moiety (myxobactin stearate, myxobactone palmitate, Figs 3.24, 3.25). The mono- cyclic aromatic carotenoids produced by *N. exedens* seem to be unique pigments (Fig. 3.26) among the Myxobacteria. The latter probably do not form glycosylated carotenoids (Reichenbach and Kleinig, 1984). Unfortunately, little has been published on the biotechnological features of these organisms. It would be very desirable to know more about the quantitative potential of Myxobacteria with regard to their carotenoid biosynthesis.

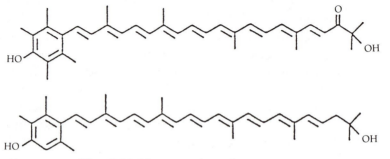

Fig. 3.26 *Nannocystis exedens* pigments.

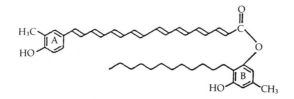

Fig. 3.27 Flexirubin.

Speaking of Myxobacteria, it is appropriate also to mention the pigments of a related group of gliding bacteria which form no fruiting bodies. The yellow or orange coloured cytophage-like bacteria (Cytophagae, Flexibacter, *Sporocytophaga spp.*, probably also certain types of Flavobacterium) produce pigments of the polyene type which have been characterized only in recent years (Achenbach *et al.*, 1974). The so-called flexirubin-type pigments have a very special structure: a σ-(4-hydroxyphenyl)-polyene carboxylic acid, esterified with a 2,5 dialkylated resorcinol (Fig. 3.27). A closely phylogenetic relationship between these bacteria has been suggested on the basis of the flexirubin-like pigments produced by a characteristic biosynthetic pathway (Reichenbach *et al.*, 1980).

Another group of bacteria in which carotenoid pigmentation has been studied more thoroughly is the Mycobacteria, especially the nontuberculous Mycobacteria, which form yellow or orange pigmented colonies. Since Mycobacteria may be photochromogenic or scotochromogenic (pigment formation in the dark) the nature and distribution of these pigments is both of physiological as well as taxonomic interest. Ichiyama *et al.* (1988) analysed 15 species and many strains, and found that all photochromogenic Mycobacteria produced β-carotene, while scotochromogenes also formed xanthophylls, both in the dark and in light. Among the xanthophylls, zeaxanthin- (*Mycobacterium phlei*) and eschscholtzxanthin- (Fig. 3.28) (*M. aureus*) like compounds or both (*M. chubuense*) were identified. They concluded that scotochromogenic Mycobacteria can be characterized by the composition of their xanthophyllic (hypophasic) carotenoids. Also, here no quantitative data are available. A biaromatic carotene (leprotene, isorenieratene, Fig. 3.29) has

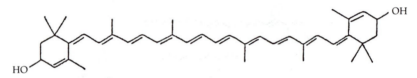

Fig. 3.28 Eschscholtzxanthin.

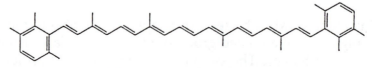

Fig. 3.29 Isorenieratene, leprotene.

been claimed to be among the carotenoids produced by *M. phlei* (Goodwin and Jamikorn, 1956). Similar aromatic carotenoids have also been described in *Streptomyces mediolanus* (Arcamone *et al.*, 1969) and more recently in *Brevibacterium linens* (Kohl *et al.*, 1983). The evolutionary and taxonomic significance of such unusual compounds should receive more attention.

Much work has been done with the pigment of the halophilic chemo-organotrophs like *Halobacterium halobium*, *H. cutirubrum* and *H. salinarum*. These bacteria live in natural salt lakes and saltern ponds and have an optimum NaCl concentration of 25% (minimum 15%). Many strains form intracellular gas vacuoles which permit floating and exposure to maximum light intensity. When grown *en masse* a reddish colour can be observed due to the carotenoid bacterioruberin, another C_{50} carotenoid (Fig. 3.22).

H. halobium cells typically swim in a straight line. When the intensity of illumination is suddenly decreased (in the red part of the spectrum) the cells stop and begin to swim in the opposite direction. This happens only if the cells are equipped with the so-called 'purple membrane' (PM), the purple fraction which is present in the intact cell as discreet patches set into and continuous with the cell membrane. It contains bacterio-rhodopsin molecules surrounded by a lipid layer attached to the inner membrane layer. The rhodopsin pigment is made of retinal and the protein opsin, and has a maximum absorption at 560–570 nm. When illuminated, the maximum shifts to

412 nm accompanied by deprotonization. The PM is formed in substantial amounts only when grown under a suboptimal concentration of oxygen and when irradiated with visible light. When growing cultures are transferred from aerobic to more anaerobic conditions, growth ceases and only the PM concentration of the cells increases sharply (Stoeckenius, 1978).

The change in pigmentation due to decreased oxygen availability, as well as the response to light intensity only when the PM is present, led many researchers to investigate the physiological function of the PM. It was found that all suspensions of *H. halobium* made anaerobic in the dark showed a marked decrease in the ATP level, which could be restored within minutes if irradiated with visible light. The action spectrum of this reversal corresponded to that of the PM. This suggested that light absorbed by the PM can generate high energy intermediates that promote ATP synthesis.

Thus, pigmented Halobacteria seem to be able to carry out a photosynthetic (ATP-forming) reaction. In contrast to all known photosynthetic organisms, they convert light to chemical energy with a rhodopsin-like pigment rather than chlorophyll. The efficiency of this reaction is probably lower than that of the chlorophylls. Also, they lack the light-harvesting apparatus of the antenna pigments found in photosynthetic organisms. Carotenoids apparently do not transfer light to the bacteriorhodopsin. Since growth under strict anaerobic conditions does not occur, it appears that the PM serves as an auxillary mechanism to satisfy the energy requirements when the oxygen concentration drops below a critical level (Stoeckenius, 1978). Why have only halophilic bacteria attempted to solve this problem of partial anaerobiosis? Is this related somehow to the halophilic environmental stress?

The fact that a structure similar to the rhodopsin of the mammalian eye was found to carry out an analogous reaction in a bacterium has caused no little stir in the scientific community.

Carotenoids in photosynthetic bacteria

Purple bacteria contain an array of carotenoid pigments that may be characterized as being aliphatic and most bear tertiary

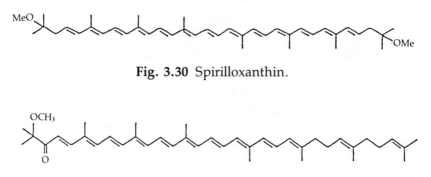

Fig. 3.30 Spirilloxanthin.

Fig. 3.31 Spheroidenone.

methoxyl groups in the 1 and/or 1′ position (Fig. 3.30). This is quite unusual in the class of naturally occurring carotenoid pigments. Most of the purple bacteria synthesize exclusively carotenoids of either the spirilloxanthin or spheroidene group (Figs 3.30, 3.31). The purple non-sulphur bacteria *Rhodospirillum rubrum* and *R. palustris,* as well as the photolithotrophic sulphur bacteria *Chromatium spp.* produce spirilloxanthin, while other purple bacteria *Rhodopseudomonas spheroides* and *R. capsulatus* produce spheroidenone, the monomethoxyl, monoketo carotenoid. *R. gelatinosa* probably produces both types of the methoxylated compounds (Sunada and Stanier, 1965).

The biosynthesis of these outstanding pigments is not much different from that of the usual C_{40} carotenoids up to neurosporene. While spirilloxanthin is derived from lycopene through a series of reactions which includes hydroxylation (rhodopin), followed by the transmethylation of methyl groups from S-adenosylmethionine, the formation of spheroidenone is carried out by another branch of the route from neurosporene, via chloroxanthin (Fig. 3.32) (not rhodopin) to the spheroidenone carotenoid (Fig. 3.33).

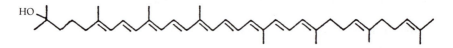

Fig. 3.32 Chloroxanthin.

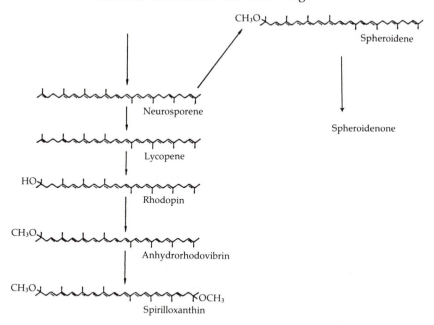

Fig. 3.33 Carotenoid synthesis in purple bacteria (simplified and adapted from Schmidt, 1978).

I shall not describe here all the carotenoids of the photosynthetic bacteria, the predominance of the methoxylated compounds, and the almost complete absence of the usual carotenoids found in higher plants, is quite impressive.

According to Liaaen-Jensen (1965) the methoxylated carotenoids are considered the most primitive carotenoids on the evolutionary scale. Although much attention has been given to the possible chemotaxonomic significance of these compounds (Liaaen-Jensen, 1979), basic questions have so far remained unanswered: what is the physiological importance of methoxylated carotenoids? Also, what can these compounds achieve, that other carotenoids can not, and why is their occurrence limited to the anaerobic photosynthetic purple bacteria? A partial answer to these questions may be derived from the study of Goedheer (1965) who investigated the fluorescence phenomenon of bacteriochlorophyll (BCl). He observed that light-induced fluorescence may be quenched (i.e. the BCl will

The carotenoid pigments

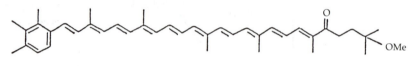

Fig. 3.34 Okenone.

be more stable) as a result of a photo-oxidative reaction resulting from energy absorption by the carotenoid pigments. From
action spectrum studies it was found that in the case of BCl
from *Rhodospirillum molishianum* and *R. rubrum*, the greatest
effect was at 566 nm, which would indicate that spirilloxanthin
is very active in quenching BCl fluorescence.

Aromatic carotenoids have also been described within the
group of purple sulphur bacteria. Okenone (Fig. 3.34) is a
carotenoid with one aromatic end group, while the aliphatic
end has a tertiary methoxyl group and a keto function in the
C−4 position. This, however, was found only in a few of the
Chromatium species (*C. oekenii, C. minus*). Aromatic carotenoids
with 1,2,5 triphenyl end groups were found in the green,
photosynthetic sulphur bacteria of the Chlorobiaceae: the
brown species, which are able to cyclize both ends (the isorenieratene series, Fig. 3.29) and the green bacteria which
cyclize only one end (the chlorobactene series, Fig. 3.35). These
compounds are probably derived from cyclic carotenes (*β*-
carotene → isorenieratene, *γ*-carotene → chlorobactene). Also,
here the chemotaxonomic significance may be emphasized,
while the functional importance of these unusual pigments still
remains obscure (Schmidt, 1978).

Erythrobacter species differ from typical photosynthetic
bacteria, being aerobic organisms that synthesize both bacteriochlorophyll and carotenoids in aerated cultures. On the other

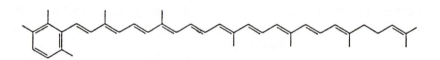

Fig. 3.35 Chlorobactene.

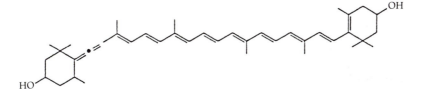

Fig. 3.36 Caloxanthin.

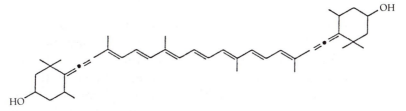

Fig. 3.37 Nostoxanthin.

hand, they cannot grow anaerobically, even if illuminated. The carotenoid population of these bacteria is therefore of special interest (Takaichi *et al.*, 1990).

Bicyclic, monocyclic and acyclic carotenoids have been described in *E. longus*, but caloxanthin and nostoxanthin (Figs 3.36, 3.37) are outstanding. The latter carotenoids do not occur in purple sulphur or nonsulphur photosynthetic bacteria, but have been described in cyanobacteria (Liaaen-Jensen, 1979) as well as in some Pseudomonas species, *P. echinoides* and *P. paucimobilis* (Jenkins *et al.*, 1979). The presence of these unusual carotenoids may somehow be related to their aerobic metabolism (see also bacteriochlorophyll in Erythrobacter, Chapter 4).

3.7 CAROTENOID PIGMENTS OF THE ALGAE

It is not necessary to describe here in detail the multitude of carotenoid compounds that have been isolated, and sometimes also quantified, in algae. Excellent reviews are now available on this subject (e.g. Goodwin, 1980). We shall, however, describe the most important carotenoids in the main groups of algae. In what way do they differ from higher plant carotenoids and to

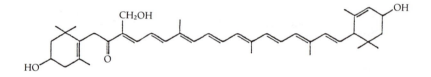

Fig. 3.38 Siphonaxanthin.

what extent may the characteristic structures encountered in algae have a chemotaxonomic or phylogenetic significance?

With very few exceptions algal carotenoids are of the C_{40} type. In the green algae (Chlorophyceae) most of the carotenoid and xanthophylls of higher plants (β-carotene, lutein, epoxy-carotenoids) can be found. However, in certain groups, especially the Siphonocladales and Codiales, the unusual xanthophyll siphonaxanthin (a hydroxymethyl group at C—19 and a keto group at C—8) (Fig. 3.38), or its ester, has been described.

While carotenoids usually occur among the chloroplast pigments, extraplastidic carotenoids have been found in certain algae cultivated under unfavourable conditions, usually with nitrogen deficiency. Cultures may appear orange or red, accumulating large amounts of β-carotene or its keto derivative echinenone (Fig. 3.39) as well as other xanthophylls. The halotolerant *Dunaliella bardawil* accumulates very large amounts of β-carotene (over 10% of its dry weight) in interplastidic globules when grown under limiting nitrogen or high salt concentrations, and exposed to high light intensity (Ben-Amotz *et al.*, 1989).

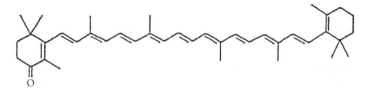

Fig. 3.39 Echinenone.

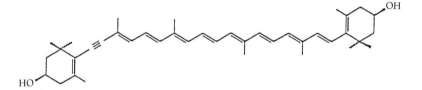

Fig. 3.40 Diatoxanthin.

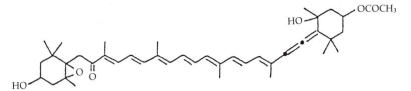

Fig. 3.41 Fucoxanthin.

Rhodophycean carotenoids do not differ much from the basic pigments described in green algae. In the xanthophycean algae, however, in addition to the basic carotenoids, the xanthophyll group often contains compounds with acetylenic bonds, for example at position 7,8 (diatoxanthin, Fig. 3.40). On the other hand the characteristic xanthophyll of the Chrysophyceae is fucoxanthin (Fig. 3.41), often accompanied by other acetylenic xanthophylls.

Similar carotenoids have also been described in Phaeophyceae and Bacillariophyceae. The brown pigment, fucoxanthin, is probably the most abundant natural carotenoid. In the Dinophyceae, a characteristic pigment, peridinin (Fig. 3.42) a nor-carotenoid, having lost three carbons in the chain, accompanied by other acetylenic carotenoids, has been found. Most

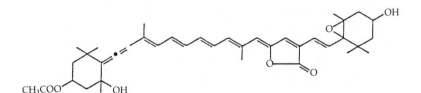

Fig. 3.42 Peridinin.

The carotenoid pigments

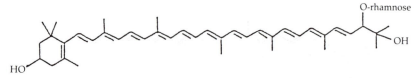

O-rhamnose

OH

HO

Fig. 3.43 Myxoxanthophyll.

of the carotenoids of the Euglenophyceae are those of the higher plant types. Even more striking is the fact that the blue–green algae usually contain the most common carotenoids: β-carotene, echinenone, zeaxanthin and the canthaxanthins, but very often also produce the unusual rhamnoside carotenoid, myxoxanthophyll (Fig. 3.43).

Recent work on the biosynthesis of carotenoid pigments in green algae has thrown new light on some of the specific reactions that take place during the formation of such pigments. Green algae are capable of performing all pigment biosyntheses when grown heterotrophically in darkness. Some mutants have been found to accumulate acyclic carotenoids in the dark but to synthesize carotenes and xanthophylls only when exposed to light. This indicates a potential light regulation during the later steps of the pathways leading to the terminal carotenoids. These later steps involve a number of enzymes (dehydrogenase, *cis-trans* isomerase and cyclase) probably located in the cytoplasmic membrane.

Mutant C—6D of the unicellular green algae *Scenedesmus obliquus* lacks the ability to form α- or β-carotene and xanthophylls when grown heterotrophically in the dark. In the absence of light, only precursors can be detected: ζ-carotene, neurosporene, lycopene and β-zeacarotene. After transfer to light the amount of precursor molecules decreases while carotene and xanthophylls (lutein) are rapidly formed.

The precursor carotenoids synthesized in dark-grown cells were found to be predominantly *cis*-isomers. Whereas *cis*-isomers of the more saturated precursors, phytoene and phytofluene, are postulated to be normal intermediates in the carotenoid pathway, *cis*-carotene, *cis*-neurosporene and *cis*-lycopene were found only in a few mutant strains. The absence of the all-*trans*-lycopene and the presence of *cis*-lycopene in the dark-grown cells of mutant C—6D may explain the inability of this mutant to form the cyclic carotenoids in the dark. Since the 3-dimensional structure of *cis*-lycopene is totally different from that of the all-*trans* forms, it is possible that the enzyme catalysing the cyclization step cannot use the *cis*-isomer of lycopene accumulated in darkness. Thus, the light regulated step would not be the enzyme catalysing the cyclization step itself but rather the isomerization of the *cis* to the all-*trans* configuration which can be followed spectrophotometrically (Humbeck, 1990).

This contention is also supported by earlier studies. When cells were illuminated in the presence of nicotine (the inhibitor of cyclization) large amounts of the *trans*-lycopene were accumulated, indicating the light-induced transformation of *cis* to all-*trans* lycopene (Davies, 1973).

Light-induced *cis-trans* isomerization may also be an important reaction in other photochromogenic organisms and should, therefore, receive more attention by current investigators.

3.8 MICROBIAL CAROTENOIDS AND THERAPEUTIC SIGNIFICANCE

In addition to the colouring attributes of microbial carotenoids in foods mentioned earlier, interest in carotenoids has recently focused on their importance in physiology and their potential use in cancer chemoprevention. Carotenoids (not xanthophylls) serve as provitamin A and are therefore essential for the process of vision in mammals. Higher animals also require vitamin A for normal growth and proper epithelial cell differentiation. However, normal physiology and well-being probably also require carotenoid function in other areas. During the last 10 years, many observations have been published with regard to the antioxidant activity of carotenoids, inhibition of mutagenesis, enhancement of the immune response, and inhibition of tumour development (e.g. Santamaria *et al.*, 1988). It is possible that these effects are somehow related, directly or indirectly, to the antioxidant effect of carotenoids. If so, which of the carotenoids are the most efficient antioxidant molecules?

Terao (1989) examined the antioxidant activity of various carotenoids in a system consisting of free radical oxidation of methyl linoleate and the formation of methyl linoleate hydroperoxides. Among the carotenoids examined in this system, canthaxanthin and astaxanthin which possess OXO groups at the 4,4' positions in the β-ionone ring retarded the hydroperoxide formation more efficiently than β-carotene and zeaxanthin which have no OXO groups. In a more recent paper, Di Mascio *et al.* (1989) confirmed the antioxidant activity of canthaxanthin and astaxanthin in a system when singlet oxygens were generated and the Kq (a quenching constant) could be measured; the Kq for the acyclic lycopene was far superior (31 for lycopene, 24 for astaxanthin).

If dietary carotenoids prove to be of therapeutic importance in the chemoprevention of cancer, one may expect, in the near future, a number of preparations for the enhancement of carotenoids in the diet. Would carotenoids such as canthaxanthin, astaxanthin and lycopene, produced by fermentation processes employing selected microbial strains, be the forthcoming source for these carotenoid pigments?

The antioxidant activity of carotenoids and their potential pharmacological use also raises another question. Little has been said so far about the configuration of these carotenoid pigments. I have mentioned earlier that, although in most studies an all-*trans* configuration is usually assumed, this has not always been critically determined. In fact, many biosynthetic studies have pointed to the occurrence also of the *cis* configuration. Modern chemical analysis employing HPLC and other techniques can differentiate between the two stereoisomers. However, little has been done so far in determining the specific antioxidant or chemoprotective activity of these isomers.

Plants and algae have been shown to contain considerable amounts of 9-*cis* β-carotene (Ashikawa *et al.*, 1986; Ben-Amotz *et al.*, 1989). From a number of studies on the nutritional effects of β-carotene stereoisomers, it seems that the 9-*cis* β-carotene absorption by chickens and rats is much higher than that of the all-*trans* isomer. This may be due to the greater solubility of the

Table 3.2 Biotechnological production of carotenoids

Source	Type	Pigment	Commercial status	Comments
Rhodotorula spp.	Yeast	Torularhodin		Undesirable hue of egg yolk
Spongiococcum excentricum	Algae	Lutein	Dried algal meal (A-Zanth, Iowa)	Discontinued
Phaffia rhodozyma	Yeast	Astaxanthin	Promising	Crustacean diet, poultry, salmonids
Anabena spp.	Cyanobacteria	Canthaxanthin		
Brevibacterium spp.	Bacteria	Canthaxanthin		
Dunaliella spp.	Algae	β-carotene	Dried algal meal (N.T.B., Western Biotechnology)	Health food
Phycomyces blakesleeanus	Fungi	β-carotene	Promising	
Blakeslea trispora	Fungi	β-carotene	Promising	

cis isomer (Ben-Amotz *et al.*, 1989). If *cis* isomers prove to be beneficial in chemoprevention of cancer, the demand for such carotenoids may increase significantly. Natural material would, of course, be the preferred source. Extracts from algae such as *Dunaliella bardawil* and *D. salina* produce comparatively high concentrations of β-carotene (up to 10% of dry weight) under suitable growth conditions. Of these, up to 60% may be the β-*cis* stereoisomer (Ben-Amotz and Avron, 1983). Considering the limitations in the cultivation of autotrophic organisms (large areas, high light intensities, harvesting) a process employing a heterotrophic organism for the production of a 9-*cis* isomer by a submerged fermentation procedure could be of great commercial interest (Shlomai *et al.*, 1991).

A summary of the potential uses and commercial status of carotenogenic micro-organisms: algae, fungi, yeast and bacteria, may be found in Table 3.2.

4

The photosynthetic pigments

It is the green pigments, the chlorophylls, that make life on our planet possible. Photosynthesis harnesses light for the process of ATP synthesis, which is the major event that leads to the fixation of CO_2 and the formation of carbohydrates.

Photosynthesis can be found in all organisms which possess any type of chlorophyll. Among the autotrophic organisms, green plants and algae employ a variety of chlorophylls which may be characterized as magnesium-containing porphyrins esterified by phytol (Figs 4.1–4.4). In the photosynthetic bacteria, the anaerobic, purple Thiorhodaceae and Athiorhodaceae,

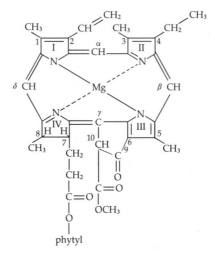

Fig. 4.1 Chlorophyll *a*.

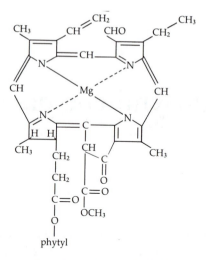

Fig. 4.2 Chlorophyll *b*.

we find bacteriochlorophylls which differ somewhat from plant chlorophylls (Fig. 4.5). On the other hand, the green bacteria (Chlorobiaceae) have the characteristic chlorobium chlorophylls which differ from normal plant chlorophylls, being esterified by farnesyl rather than by phytol, and containing some other

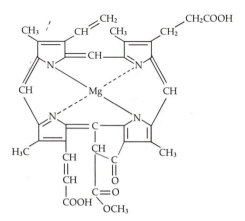

Fig. 4.3 Chlorophyll *c*.

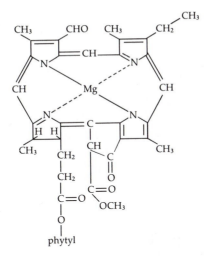

Fig. 4.4 Chorophyll *d*.

substituents at various carbon positions of the porphyrin structure (Figs 4.6, 4.7).

It is in common usage to distinguish between oxygenic and anoxygenic photosynthesis. In the former, oxygen is evolved during the process of photosynthesis, as occurs in green plants,

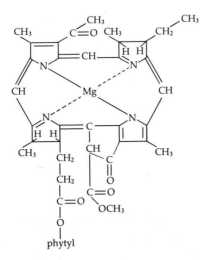

Fig. 4.5 Bacteriochlorophyll *a*.

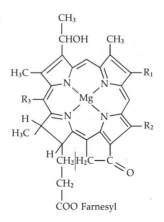

Fig. 4.6 Bacteriochlorophylls *c* and *d* (chlorobium chlorophylls).
R_1 = isobutyl, *n*-propyl, or ethyl,
R_2 = ethyl or methyl,
R_3 = methyl.

algae and the blue-greens (Cyanobacteria). The anoxygenic photosynthesis is thus limited to the purple and green bacteria.

Individuality of light absorption spectra is one of the major characteristics of the chlorophylls. While plant chlorophylls

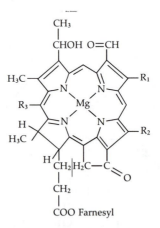

Fig. 4.7 Bacteriochlorophyll *e* (chlorobium chlorophyll).
R_1 = isobutyl, *n*-propyl, or ethyl,
R_2 = ethyl or methyl,
R_3 = methyl.

strongly absorb at around 450 nm and 650 nm, bacteriochlorophylls absorb light in the near infra-red region (660–870 nm).
This, as we shall see, is of great ecological importance. Another
typical feature of bacteriochlorophylls is that they are not contained in chloroplasts (like plants and algae), but may be found
in extensive membrane systems that occur in the bacterial cell.

Photosynthesis is usually considered to comprise two
reactions: the build-up of high-energy molecules (ATP) and
reducing power (NADPH) which are light dependent, and the
reduction of CO_2 and the formation of carbohydrates, employing ATP and reducing power, which are light independent (the
dark reaction).

Photosynthesis in bacteria starts with the excitation of the
chlorophyll molecule by light quanta. Excitation of the bacteriochlorophyll results in an electron being driven off, and
the molecule becomes positively charged. The electron is
transferred to ferredoxin down an electrochemical gradient,
through a chain of electron carriers (a series of cytochromes),
and is returned to the positively charged bacteriochlorophyll
(Fig. 4.8). During the passage of the electron between cytochromes *b* and *f* the high energy ATP is formed from ADP and
inorganic phosphate. As the electron is completing a cycle

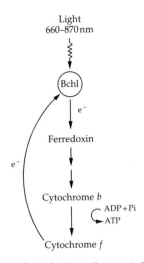

Fig. 4.8 Cyclic phosphorylation (bacterial photosynthesis).

82   *The photosynthetic pigments*

upon its return to the bacteriochlorophyll this process is known as cyclic photophosphorylation. Thus, bacterial photosynthesis provides only ATP, but no reducing power. The latter is accomplished by the reducing activity of other compounds in the environment, such as H_2S or other organic molecules.

In plants and algae another system is operative. This comprises two phases: photosystem I (PSI) where chlorophyll is excited by red light, and photosystem II (PSII), somewhat reminiscent of the bacterial system, where the chlorophylls are excited mostly by blue light. In PSII, photolysis of water occurs producing O_2 (oxygenic photosynthesis) and two electrons, which are transferred from the hydroxyl ions of water through the chain of electron carriers, yielding ATP. The two electrons continue their passage to PSI, boosted by red light they again form ATP, eventually reaching $NADP^+$, where reducing power in the form of NADPH is formed. Hence, the electrons, unlike in the case of bacteriochlorophylls, are not returned to the original chlorophylls. This is known as non-cyclic photophosphorylation (Fig. 4.9). Although many details of the photosynthetic process are still unclear the general layout of the reactions

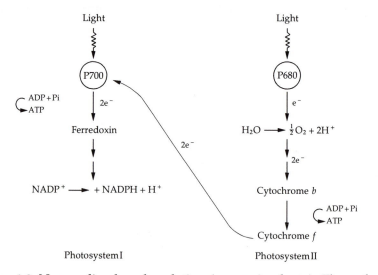

Fig. 4.9 Non-cyclic phosphorylation (oxygenic plants). Flow of electrons during phosphorylation (simplified).

described clearly distinguishes between bacterial photosynthesis and oxygenic systems. Another point, which has been frequently made in the older literature with regard to the analogy of bacterial and plant photosynthesis, was formulated by Van-Niel:

$$2H_2O + CO_2 \rightarrow (CH_2O)_x + O_2 + H_2O$$

$$2H_2A + CO_2 \rightarrow (CH_2O)_x + 2A + H_2O$$

$$2H_2S + CO_2 \rightarrow (CH_2O)_x + 2S + H_2O$$

It is no longer tenable, since the reducing hydrogen atoms are not involved in the photochemical reactions that lead to ATP formation.

The chlorophyll (Chl) population is not homogenous but differs among various organisms, and apparently also in various sites within the thylakoids, the membranous, flattened vesicles of the chromatophores. There are two main locations: the reaction centre (RC) where the photochemical reaction takes place and the 'light harvesting' (LH) location where light quanta are absorbed and transferred to the RC. LH is populated by the so-called 'antenna pigments'. In green plants and algae, the antenna pigments consist of a number of protein complexes of Chl *a* (Fig. 4.1), Chl *b* (Fig. 4.2) and various carotenoids. The light energy trapped by these pigments is transferred to the RC which contains Chl *a*. There is a slight difference in the chemical structure between Chl *a* and Chl *b*. A methyl group at C3 (ring II) of Chl *a* is replaced by an aldehyde (CHO) in Chl *b*. This gives a slightly different absorption spectrum (peaks at 435 nm and 645 nm versus 425 nm and 660 nm of Chl *a*). In green plants the ratio $a : b = 2 : 1$. The chlorophyll content of green plants is much higher than that of algae (6–8% of dry weight in spinach leaves in comparison with 0.3–2% in algal cells).

Chl *b* is also found in the 'green plant line' of algae (Chlorophyceae, Prasinophyceae and Euglenophyceae). The Chl *a* : Chl *b* ratio is also similar, ranging from 2 : 1 to 3 : 1. In the 'brown plant line' (Cryptophyceae, Dinophyceae, Chrysophyceae, Phaeophyceae, Xanthophyceae, Bacillariophyceae, Haplophyceae, Raphidophyceae) there is no Chl *b*, but Chl *c* (Fig. 4.3) may function as an accessory pigment. This chlorophyll is characterized by an unesterified $CH=CH-COOH$ group at C7 (IV) and a double bond at C7,8 (IV). Actually there are 2 types of Chl c: $c1$ and $c2$, the latter having a CH_2-CH_2-COOH at C4 (II). The ratio Chl a : Chl c varies from 1.2 : 1 to 5.5 : 1. For a more detailed description of the chlorophyll pigments the interested reader should consult Goodwin's treatise on the *Chemistry and Biochemistry of Plant Pigments* (1976).

In the red algae (Rhodophyceae) another chlorophyll may be found, Chl d (Fig. 4.4), having a CHO group at C2 (I) with a peak at somewhat longer wavelength (696 nm). In the blue-greens (Cyanophyceae) and some Rhodophyceae, Xanthophyceae and Chrysophyceae, no accessory chlorophylls have been found.

Although the biosynthesis of algal chlorophylls follows the pathways employed by higher plants, there is a major difference with regard to the last step, the conversion of protochloro-phyllide (PCHL) to chlorophyllide, the phytolless precursor of chlorophyll. This reductive reaction is light dependent in higher plants. Hence, plants grown in the dark accumulate PCHL (Fig. 4.10). In algae, with some exceptions, grown heterotrophically in the dark, the conversion to chlorophyllide and chlorophyll takes place in the absence of light.

Glancing at the absorption spectra of the pigments involved in photosynthesis (the chlorophylls and carotenes) one is

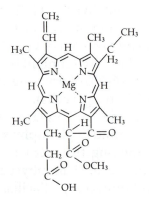

Fig. 4.10 Protochlorophyllide *a*.

struck by the poor absorbance in the region of 500–600 nm. Roughly speaking this would constitute about 30% of the visible range which is little used by plants containing these pigments. However, nature has provided a number of other pigments that absorb strongly in this range.

Many of the algae which do not contain Chl *b* (blue-greens, red algae and cryptomonads) form extrathylakoidal accessory pigments, the phycobiliproteins. Aggregates of these proteins form the so-called phycobilisomes, which are attached to the outer surface of the photosynthetic lamellae. These pigmented proteins are very active in the light harvesting process and may contribute up to 50% of the LH capacity of these cells.

Phycobiliproteins may be characterized by their prosthetic groups which are responsible for their specific absorption capacity. This is due to the open chain tetrapyrrole structures that are covalently bound to these proteins. The chemical structures of the major 'blue' chromophore, phycocyanobilin (PCB) and red chromophore, phycoerythrobilin (PEB) are known (Figs 4.11, 4.12). Cyanophyte and rhodophyte phycobiliproteins contain either PEB or PCB, or both, as their prosthetic

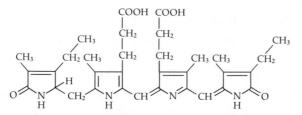

Fig. 4.11 Phycoerythrobilin (bilin chromophore).

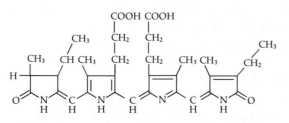

Fig. 4.12 Phycocyanobilin (bilin chromophore).

groups. The terms phycoerythrin and phycocyanin are usually employed to designate phycoerythrobilin- and phycocyano-bilin-containing phycobiliproteins.

Algal biliproteins have been studied since the early 19th century. The brilliant, attractive colouration and fluorescence of biliprotein solutions may have been a major feature in their attraction to early workers. They are globular proteins and may be readily extracted from ruptured cells into water or dilute salt solutions, and separated by conventional salting out procedures. Phycoerythrin (max. at 570 nm) assumes a clear red appearance by transmitted light and emits a brilliant orange-yellow fluorescence. The phycocyanins (max. 620 nm) are blue with a strong red fluorescence (Carra and hEocha, 1976). Most species contain both biliproteins, although a single biliprotein usually predominates. Generally, the phycoery-thrins are the most abundant in the red algae, while phyco-cyanins constitute most of the blue-green phycobiliproteins. There are some exceptions, for example in some, blue-green phycoerythrin predominates, resulting paradoxically in 'red' blue-greens and *vice versa* in some red algae; the unicellular *Porphidium aerugineum* is blue, having only phycocyanin as phycobiliprotein.

The phycobiliproteins may constitute a major portion of the algal proteins. Up to 40% of cell proteins, or 24% of the dry matter of blue-greens have been reported. This permits efficient light trapping, even at low light intensities. The ratio between phycoerythrin and phycocyanin may vary according to growth conditions and light regimes. Nitrogen deficiency has been shown to cause a selective loss in phycobiliprotein content in *Pseudoanabena sp.* (Canto de Loura *et al.*, 1987).

Contrary to the carotenoid complexes of the antenna pigments, which can be removed with organic solvents, the covalent attachment of the phycobilins to their respective apoproteins makes their extraction more difficult, requiring more severe lytic conditions.

The algal biliproteins act as accessory photosynthetic pigments, absorbing very efficiently in the spectral regions in which chlorophyll absorbs only poorly or not at all. Differential scattering and absorption of both the blue and red light ends of the spectrum by sea water, results in greater penetration through deep water of light in the mid-spectral region (green-yellow), which phycoerythrin absorbs efficiently. Indeed, marine red algae are relatively richer in phycoerythrin than those grown in the littoral zone, although several cases have been reported of shallow water algae which were also rich in phycoerythrin.

Another phycobiliprotein which usually occurs at very low concentrations is allophycocyanin (max. at 650 nm). This pigment protein is considered to function as an energy funnel, located between lamellar Chl a and other phycobilisomes. Energy harvested by the phycobilisomes is thus transmitted through allophycocyanin to the reaction centre.

Phycobiliproteins may differ in a number of parameters: the subunit structure of the apoprotein usually consists of two dissimilar polypeptide chains (α, β), and sometimes three chains (α, β, γ); each subunit may carry one, two or four prosthetic groups of different bilin types; and the phycobiliproteins may assume different stages of aggregation, reaching high molecular weights (e.g. $(\alpha, \beta)_6$). All these variations affect the specific absorption spectra of the pigment proteins.

In addition to the prosthetic groups mentioned so far, there seem to occur additional bilins, such as phycourobilin (PUB) which endows an additional peak (498 nm) in the absorption spectrum of some phycoerythrins. This has been found in marine *Oscillatoria irrigua*, *Trichodesmium thiebantii* and in fresh water *Gloebacter violaceus* (Stadnichuk *et al.*, 1985). The structure of PUB is still unknown.

Various biophysical considerations seem to indicate that the biliproteins of blue-green and red algae share a common evolutionary origin. This does not apply to the Cryptomonads, which are phylogenetically distant from the Rhodophyta and have biliproteins which are immunologically distinct.

Phycobiliprotein colour and fluorescence are stable over a wide pH range and can be stored for long periods of time. They are readily soluble in cold to warm water, yielding a uniform, clear aqueous solution. They are also soluble in ethanolic solutions of less than 20%. Because of these features, research is now being conducted by a number of commercial enterprises to use these pigments for colouration in cosmetic preparations.

Interestingly, the open chain tetrapyrrole structure can also be found in higher plants in the form of phytochrome, the light sensory probe that controls flowering in many plants. The prosthetic group of phytochrome is very similar to phyco-cyanobilin, but no light harvesting function has been ascribed to such pigments in higher plants.

4.1 BACTERIAL PHOTOSYNTHETIC PIGMENTS

The non-oxygenic photophosphorylation in bacterial autotrophs involves a number of chlorophylls and carotenoids that differ somewhat from the pigments of green algae and higher plants. Most species of purple bacteria contain bacteriochlorophyll (Bchl) *a* as their only bacteriochlorophyll but some, like *Rhodo-pseudomonas viridis* and *Ectothiorhodospira halochloris* contain Bchl *b* instead. Bacteriochlorophylls differ from plant Chls by the absorption at longer wavelengths (up to 1020 nm by Bchl *b*). The Bchls of the green sulphur bacteria are outstanding in that most of their chlorophylls have a farnesyl residue that replaces the phytol moiety of the common chlorophylls.

Green bacteria

These comprise two families: the Chlorobiaceae, and the Chloroflexaceae. The former are strict photolithotrophs and obligate anaerobes, while the latter are photo-organotrophs and facultative anaerobes. Most of the antenna pigments of these organisms are located in chlorosomes (formerly chloro-bium vesicles). These are bag-like structures appressed to the cytoplasmic membrane, whereas the RCs are located within the cytoplasmic membrane. The LH apparatus contains mainly Bchl *c* and some carotenoids. All species have Bchl *a* in the RC. In *Chlorobium limicola* and *Prosterochloris aestuari* (Chloro-biaceae) the photosynthetic apparatus contains an abundance of molecules of Bchl *c*, *d* or *e* in the LH, versus only a few molecules of Bchl *a* in the RC. This enormous LH system enables the green bacteria to grow abundantly in weak light. The Bchl content of green bacteria is very sensitive to environmental changes and varies inversely with light intensity. *Chloroflexus aurantiaca* grown at 240 lx contains 40× more Bchl *c* than cells grown at 52 000 lx. This is apparently due to a change in the number of chlorosomes (Pierson and Castenholtz, 1978).

Purple bacteria

Contrary to the green sulphur bacteria, the photosynthetic LH apparatus of the purple, sulphur and non-sulphur bacteria, is not contained within bag-like structures but is located within the intracytoplasmic membrane system which is continuous with the cytoplasmic membrane of the cell. These bacteria form Bchl *a* and occasionally Bchl *b*. In *Rhodopseudomonas viridis*, the Bchl *b* has an unsaturated substituent at position 4 which causes a pronounced shift of the intense absorption band in the far-red region to longer wavelengths (max. 1015–1034 nm). In terms of LH efficiency these bacteria are less advanced, the ratio between antenna pigment molecules and RC being only 60 in comparison to 2000 of the green bacteria (Chlorobiaceae). In higher plants this ratio is 200–300.

Aerobic photosynthetic bacteria

Contrary to the characteristics of the classical photosynthetic bacteria (described above) which thrive under anaerobic conditions when illuminated, Shiba and Harashima (1986) described a new bacterium, *Erythrobacter sp. OCh 114* which can be induced to form Bchl under aerobic conditions. They do not grow anaerobically in light or darkness and have been referred to as aerobic photosynthetic bacteria. Bright light was shown to suppress Bchl biosynthesis almost completely. It was shown that aerobic growth in darkness leads to the formation of Bchl which begins to function photosynthetically upon exposure to light with ample oxygen supply. The photosynthetic electron transport system in this bacterium cannot replace the respiratory electron transport system in response to anaerobiosis.

One of the intriguing questions with regard to microbial and plant photosynthesis is the architecture of the photosynthetic apparatus. Being outside the scope of this review, the interested reader will find much material on this subject in the comprehensive treatment by Glazer (1982).

4.2 BIOSYNTHESIS

The porphyrin structure of chlorophyll is very similar to other life-sustaining tetrapyrrole pigments such as haem and vitamin B_{12}. In spite of some differences in structure and metal content (Mg^{2+} in chlorophyll, Fe^{2+} in haem, Co^{2+} in vitamin B_{12}) resulting in different colours and physiological functions, the biosynthesis of these molecules follows a similar pattern. Much of our knowledge in this area has become available from the brilliant work of the Cambridge group headed by Professor Battersby (1988).

5-aminolevulinic acid produces the monopyrrole porphobilinogen (PBG). Four molecules of PBG form the central precursor of the porphyrin structure, uroporphyrinogen III (uro'gen III) by the action of two enzymes, a deaminase and a cosynthetase (Fig. 4.13). Uro'gen III is not a pigment (the pyrrole rings are joined by methylene bridges, not a conjugated system) nor does it complex with metal ions. The assembly of the open tetrapyrrole structure (hydroxymethylbilane) is carried out by the deaminase enzyme with the release of ammonia. This is followed by the closure to uro'gen III by the cosynthetase enzyme which is also responsible for some rearrangement in ring IV. The transformation of uro'gen III to coproporphyrinogen III and protoporphyrinogen IX involves an oxidative decarboxylase system in eucaryotes. Since photosynthetic bacteria produce Bchl under anaerobic conditions, this step must differ somewhat from the oxidative counterpart. Protoporphyrinogen IX is then transformed into protoporphyrin IX by dehydrogenation, leading to the double bonds in the interannular bridges. Protoporphyrin IX is the branch point

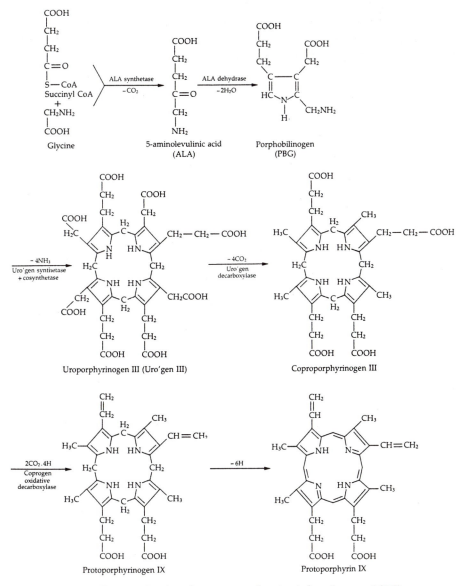

Fig. 4.13 Biosynthesis of protoporphyrin (after Jones, 1978).

where Fe^{2+} insertion leads to haem formation (cytochromes) and Mg^{2+} insertion results in the Mg-protoporphyrin that leads to Bchl. Similar to many algae also, bacterial chlorophylls may be synthesized in the dark-grown heterotrophic photosynthetic bacteria which are capable of chlorophyll synthesis in the absence of light (Jones, 1978).

5

The photosensitizing pigments

Many plants are known to contain a variety of toxic compounds belonging to different chemical categories. Photoactive chemicals that become active only under the influence of light, the photosensitizing compounds, constitute a special category.

An interesting case in which a pigment of the tetrapyrrol type is involved in a process designated as photosensitization, resulting in severe damage to animals, has been described. A South African sheep disease known as 'geeldikkop' (yellow, thick head) in which jaundice leading to a yellow colour and swelling of the skin of the head was found to be the result of an interesting combination of nutritional factors and microbial activity. Chlorophyll made available during the digestion of green plant material by ruminant flora is transformed in part to phylloerythrine (structure undefined). This is normally voided from the animal by way of the bile into the faeces. Under some conditions, which are not fully understood, phylloerythrin enters the blood stream and reaches the skin, where it can start the process of photosensitization upon exposure of the animal to sunlight, resulting in the condition described.

Photosensitization is, thus, a process in which a particular compound absorbs some radiation and transmits the absorbed energy to certain sensitive sites in the tissues, which then leads to a specific effect or disease. In many cases the photosensitizing molecule need not be a colourful pigment, as conjugated double bonds are quite efficient in the absorption of ultraviolet light. An example is the psoralens which belong to the group of

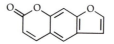

Fig. 5.1 Psoralen.

furocoumarins (Fig. 5.1). These compounds are quite abundant in some of the plant families (Umbelliferae, Leguminosae) and are natural photosensitizers that may cause dermatitis, erythema, blisters, itching or scaling of the skin after exposure to light. In some cases, gastric irritation, nausea, nervousness and even depression have been reported. Various bacteria seem to have a different sensitivity to the psoralens, the Gram positives have been found to be more susceptible than Gram negatives. A correlation between radiation resistance and resistance to photosensitization by psoralens has been suggested. Since psoralens lead to a striking increase in the number of penicillin resistant mutants, they have been considered as mutagenic compounds acting upon bacterial DNA (Mathews, 1963).

Photosensitizing pigments are quite common in some plants. The best known example is that of hypericin (various *Hypericum* species). Grazing animals, such as white sheep and horses, who consume these plants may become vulnerable. In other animals only the unpigmented portions of the skin show irritation. Glass filtered light was shown to be effective in the induction of hypericism. A highly fluorescent red pigment has been isolated from these plants (Fig. 5.2). Affected areas of the skin redden and show blisters like sunburn, which may

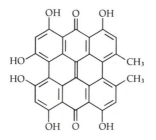

Fig. 5.2 Hypericin.

become infected. Tissues may become oedematous and fluids accumulate subcutaneously, resulting in the swelling of various parts of the body. Necrotic areas will not bear wool.

Hypericin-like molecules have also been described in microbial organisms. The fungus *Penicilliopsis clavariaeformis* has been reported to produce such a pigment (Brockman and Eggers, 1958). A hypericin-like molecule has also been described in the Basidiomycete *Dermocybe austroveneta*, a common mushroom in Australian eucalyptus goodlands (Gill *et al.*, 1988). When sporophores of *D. austroveneta* are attacked by insects or suffer damage or ageing, the flesh assumes a red-violet colouration which, upon extraction, yields a deep red-violet solution with intense red fluorescence.

The most interesting and best studied case is that of the protozoan Blepharisma. Blepharismas are single-celled spirotrich ciliates and may usually be found as bottom dwellers in ponds and waterways. A number of species are known, differing in size (60–300 μm), shape and colour. Some of these differences are certainly also due to particular environmental conditions. Blepharismas feed on bacteria but can also be grown in axenic culture. They may become carnivorous if small ciliates are present or even cannibalistic, eating smaller individuals of their own species when no other food is available. Their size depends a great deal upon their mode of nutrition. Being bottom dwellers they are usually shaded by debris and assume a pink to red colour. Upon prolonged illumination with dim light (200 fc) they will turn greyish blue to colourless. However, exposure to strong light (2700 fc) at 550 nm and below, in the presence of oxygen, will kill them. This does not happen when the ciliates are first bleached by dim light or in the absence of oxygen. This is a very remarkable photosensitizing effect. The pigment may easily be obtained from the ciliate. After chilling for 30–120 s at 0 °C, Blepharismas shed the pigment, whereas upon return to normal temperature in darkness the red pigment will form again if suitable nutrients are available. The ciliate may, thus, be 'milked' a number of times. The water soluble pigment occurs in arrays of symmetrically arranged granules under the pellicle, between two rows of cilia. Each granule is about 0.75 μm thick and is bounded by a very thin membrane. The pigment has been named blepharismin and is related structurally, but not identically, to hypericin.

The red-fluorescent pigment is toxic in concentrated solution to various colourless Protozoans, even in darkness. However, upon dilution it becomes innocuous in darkness but induces sensitivity when exposed to bright light. The photosensitization process can be best observed at 8 °C, when illuminated pink cultures will not expel water but become bloated and eventually burst.

Why is an organism producing a molecule that, under specific circumstances, will lead to its death? The strong absorption of blepharismin in the far ultraviolet region (200 nm) and the location of the pigment granules in the outer layer of the cell seem to indicate a protective function in the filtering out of this lethal radiation. If life on earth is possible because the ozone layer does not permit such lethal radiation to reach the earth, it is possible that Blepharismas are some kind of living fossils of ancient times, before ozone was formed (Geis, 1981). This contention receives support from the fact that colourless Blepharismas are much more sensitive to short ultraviolet light than the pigmented cells. However, this explanation does not cover the fact that this very pigment carries the photosensitizing character that makes light lethal. Nature could have done better!

6

Flavinogenic microbes

The most important flavin compound produced by micro-organisms is riboflavin (Fig. 6.1). Although an important vitamin (B_2) with many interesting physiological, biochemical and economic features, it is rarely considered or used as a pigment. And yet, a neutral aqueous solution of riboflavin has a greenish-yellow colour (with maxima at 311, 444 and 475 nm) and an intense yellowish-green fluorescence (max. at 530 nm). Being water soluble and produced in considerable quantities by a variety of micro-organisms, considering this compound among microbial pigments is certainly justified.

Many micro-organisms, fungi and yeasts, as well as certain bacteria, synthesize and excrete riboflavin into their growth media. While riboflavin for pharmaceutical purposes is so far produced by chemical synthesis, the supply of this important

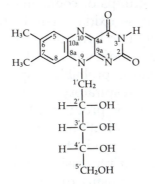

Fig. 6.1 Riboflavin.

vitamin in feedstock for animals is usually in the form of crude concentrates derived from the fermentation industry where selected strains of a number of organisms have been shown to produce 6–7 g/l of the desired compound.

The organisms employed for the biosynthesis and production of riboflavin may be divided into two categories: those not affected by iron, and those which are very iron sensitive. Since most of the production work is done with complex media containing various industrial, not refined, raw materials, the insenstivity to iron ions is of great importance. Among this group two Ascomycetes, yeast-like plant pathogenic fungi have been found to be most suitable: *Eremothecium ashbyii* and *Ashbya gossypii*. Good yields have been obtained when grown on cornsteep liquor and corn oil. *A. gossypii* has a certain advantage over *E. ashbyii* in that the latter readily gives rise to stable non-flavogenic strains. Yields with yeasts (e.g. *Candida*, *Pichia* and *Hansenula spp.*) as well as bacteria (*Clostridium*, *Bacillus* and *Brevibacterium spp.*) were usually much inferior.

A number of patents have been issued employing some of these organisms on specific media such as ethanol, methanol or paraffins (Yoneda, 1984).

6.1 PHYSIOLOGY

In biological systems, riboflavin functions almost exclusively in the form of flavoproteins, in which FMN or FAD are generally bound as prosthetic groups or coenzymes to specific proteins. These enzymes catalyse oxidation–reduction reactions. Riboflavin readily takes up two hydrogen atoms to form the almost colourless 1,3-dihydroriboflavin, which may be reoxidized by air. This redox system is probably responsible for its physiological function in the respiratory chains.

Many micro-organisms produce riboflavin, although normally intracellular concentrations are quite low. The mechanism by which the rate of production is kept low is not known, but seems to be due to repression and low rate of enzyme formation. Even in organisms that produce and excrete considerable quantities into the medium, the internal riboflavin concentration is very low, most of the flavins appearing as mono- or dinucleotides (FMN, FAD). However, overproducers of riboflavin like *A. gossypii* may show higher internal concentrations, although the internal nucleotides are at a normal level. Hence, overproduction must be somewhat related to a lack of repression in the formation of the flavin itself.

Ascospore formation by sporogenous strains of the yeast-like fungi occurs in the overproduction stage and it has been suggested that some correlation must exist between sporulation and riboflavin production. *A. gossypii* strains which are poor riboflavin producers also show poor sporulation. In good producers, heavy sporulation takes place. The site of riboflavin crystal accumulation is found to be within immature asci. However it has been noted that, in complex or defined fermentation media with good riboflavin yields, sporulation rarely takes place. It is possible that potentially good riboflavin producers show good sporulation only on solid media.

6.2 BIOSYNTHESIS

Purines stimulate riboflavin production in *Candida spp.* and *A. gossypii*. Glycine, a common purine precursor, acts similarly. Both seem to stimulate riboflavin biosynthesis directly and not via growth promotion. This was shown in stationary, not growing, cultures. The two heterocyclic rings of riboflavin are indicative of a purine origin. Numerous workers have studied the identity of the immediate purine precursor employing various labelled compounds and auxotrophic mutants. It has been concluded that all purines must be converted to guanine (nucleotide) before incorporation. It is not entirely clear whether the ribose of the nucleotide precursor is also incorporated or whether the pentose is derived from another source. The guanine molecule is transformed to a pyridine (6-hydroxy 2,4,5-triaminopyrimidine-HTP). After incorporation of the ribityl moiety at the C4 amino group, a deamination takes place at C2, and a C4 (acetoin?) compound is incorporated, building the central ring of the final product. This compound, a lumazine designated as DMRL, is then converted to riboflavin by a

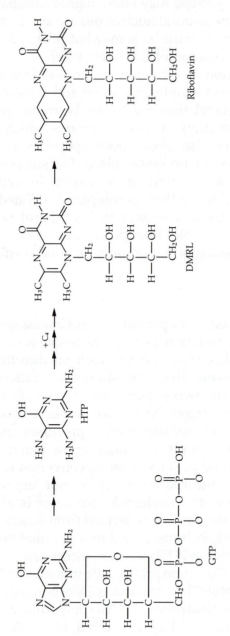

Fig. 6.2 Riboflavin biosynthesis (simplified, after Demain, 1972).
GTP = guanosine triphosphate,
HTP = 6-hydroxy 2,4,5-triaminopyridine,
DMRL = 6,7-dimethyl-8-ribityl-lumazine.

mechanism in which one molecule of DMRL donates its recently acquired C4 moiety to a second DMRL molecule, yielding riboflavin (Fig. 6.2).

The sensitvity of most organisms to iron ions and the insensitivity of the yeast-like overproducers is a very interesting phenomenon which demands an explanation. It has been suggested that iron stimulates the iron-containing enzymes xanthine oxidase and urate oxidase, shunting purines away from the riboflavin pathway. However, mutants lacking xanthine oxidase synthesized riboflavin in iron-rich media at the same rate as the wild type. In the case of the yeast-like overproducers it is thought that an iron-containing flavoprotein regulates the biosynthesis of riboflavin as an enzyme repressor only under very specific environmental conditions. *A. gossypii* grows abundantly over a wide temperature range, while riboflavin production is limited to a narrow range below the optimum of growth (26–28 °C). It is suggested that riboflavin overproduction is controlled by a repressor that is not formed during growth at low temperatures.

Most of the work on the biosynthetic pathway described above has been pursued during the 1950s and 1960s. I have therefore omitted all pertaining references. The interested reader should turn to the review article by Demain (1972) for a complete picture of riboflavin biosynthesis and its regulation.

7

Phenazine pigments

Among the pigments produced by micro-organisms, the phenazine pigments (Fig. 7.1) have a number of outstanding features. Their production is limited to certain bacterial genera and they may be either highly or sparingly soluble in water. While pyocyanine (Fig. 7.2) causes the blue colourization of the growth medium of *Pseudomonas aeruginosa*, chlororaphine (Fig. 7.3) and iodinin (Fig. 7.4) produced by *Ps. chlororaphis* and *Brevibacterium iodinum*, respectively, crystallize out in emerald green and dark purple crystals within or around their colonies, and yet their chemical structures are very similar. Also *Ps. aureofaciens* deposits golden yellow crystals of similar structure. Furthermore, the formation of the phenazine pigments is not

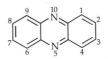

Fig. 7.1 General structure of phenazine compounds.

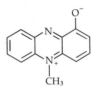

Fig. 7.2 Pyocyanine.

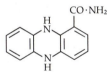

Fig. 7.3 Chlororaphine.

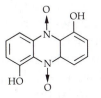

Fig. 7.4 Iodinin.

limited, as initially thought, to the genus Pseudomonas, but may also be found in other Gram negative and Gram positive bacteria.

Research on phenazine pigments has been stimulated not only by their appealing colours and fairly easy production in ordinary laboratory facilities, but also because of certain features which suggest some potential medical interest.

Over 50 naturally occurring phenazine compounds are presently known. Pyocyanine is probably the best known. It was described over a century ago on infected wound dressings. These pigments are produced by several organisms of the microbial population of the skin of humans and animals. Chromogenic strains of *Ps. aeruginosa* are potent pigment producers and may stain infected wounds with a blue colour. The indicator characters (acid-base, and redox) were earlier described. The 'chameleon phenomenon', the changing of colour due to the disturbance of cultures growing on solid media, was soon recognized to be due to the opposing action of atmospheric oxygen and the reducing power of the bacteria. Pyocyanine must not be confused with the green fluorescent pigment (pyoverdin) also produced by various Pseudomonas strains. This fluorescent pigment is now known to be a siderophore (Meyer and Abdallah, 1978).

Production of pyocyanine is fairly simple. A medium containing glycerol, leucine, glycine or alanine and mineral salts was
found suitable. An inhibitory effect of phosphate on pyocyanine
formation has also been noted. Pyocyanine production is typical
of a secondary metabolite, with its biosynthesis taking place
after growth has ceased (Halpern *et al.*, 1962). Different strains of
Ps. aeruginosa produce a variety of different phenazine compounds such as oxychlororaphine (phenazine-1-carboxamide),
chlororaphine and phenazine-1-carboxylic acid (Figs 7.5, 7.6).
Conditions have been studied for the preferential production of
specific phenazine compounds. Pyorubin, another phenazine
molecule, is a bright red water soluble pigment which has been
found in a number of fresh isolates of *Ps. aeruginosa*. Actually,
pyorubin is a mixture of two phenazines (aeruginosin A and B)
(Figs 7.7, 7.8). Some strains produce pyocyanine as well as
pyorubin.

The yellow pigment, phenazine-1-carboxylic acid (Fig. 7.6)
can be produced by *Ps. aureofaciens* when grown on a synthetic
medium containing gluconate and mineral salts, along with a
number of various phenazine derivatives. Green needle-shaped
crystals are produced by old cultures of *Ps. chlororaphis*. Also,
the chlororaphine pigment has turned out to be a mixture of
phenazine-1-carboxamide (oxychlororaphine, Fig. 7.5) and its
5,10-dihydro derivative, chlororaphine (Fig. 7.3). For production, a medium containing glycerol, peptone and mineral salts

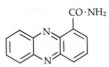

Fig. 7.5 Oxychlororaphine (phenazine-1-carboxamide).

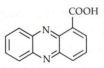

Fig. 7.6 Phenazine-1-carboxylic acid.

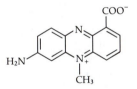

Fig. 7.7 Aeruginosin A.

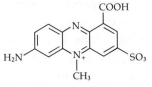

Fig. 7.8 Aeruginosin B.

has been found useful. While pyocyanine-producing *Ps. aeruginosa* may also produce oxychlororaphine, none of the *Ps. chlororaphis* strains examined will produce pyocyanine.

A soil Pseudomonad identified as *Ps. cepacia* was found to produce a purple pigment of the phenazine family (4,9-dihydroxyphenazine-1,6-dicarboxylic acid dimethyl ester, Fig. 7.9) when cultivated on a medium containing glycerol, peptone and mineral salts. This was usually accompanied by other phenazine derivatives. Another purple phenazine pigment (iodinin, Fig. 7.4) has been described in a culture of *Ps. phenazinium* when grown on a threonine or glycine mineral salts medium. Media containing glucose or glycerol were repressive for iodinin production. The taxonomic position of this organism has not been defined since it has been described showing peritrichous flagellae and other features inconsistent with this genus.

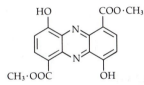

Fig. 7.9 4,9-dihydroxyphenazine-1,6-dicarboxylic acid dimethyl ester.

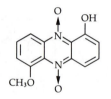

Fig. 7.10 Myxin.

Phenazine compounds, predominantly iodinin and 1,6-dihydroxy-phenazine have been described in a number of Actinomycetales: *Brevibacterium*, *Corynebacterium*, *Streptosporangium*, *Nocardia*, *Microbispora* and a variety of *Streptomyces spp.* Phenazine compounds can be encountered quite frequently among the Actinomycetales. A number of Myxobacteria (e.g. Sorangium) have also been found to synthesize some phenazines. A cherry-red pigment (myxin) which is very similar to iodinin has been described in a Sorangium strain (Hanson, 1968).

Many of the phenazine compounds show antibacterial activity. They were probably the first bacterial metabolites to show antibiotic activity against other organisms. Indeed many of the antibiotic compounds isolated and described in fermentation liquids of the Actinomycetales (e.g. griseolutein) were found to be phenazine derivatives or contain the phenazine nucleus.

The planar aromatic phenazine structure is reminiscent of many compounds (acridine, actinomycin, etc.) that are known to be intercalating agents with DNA. Indeed phenazines were found to be inhibitory to DNA-dependent RNA synthesis. However, other systems may also be involved in the phenazine inhibition. Baron and Rowe (1981) have shown that denitrifying

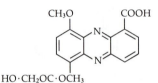

Fig. 7.11 Griseolutein A.

bacteria were more susceptible during anaerobic than aerobic cultivation when exposed to pyocyanine. On the other hand, *E. coli* inhibited by phenazines showed enhanced oxygen uptake which may be connected with the production of superoxide radicals and H_2O_2.

Few phenazine compounds have been released for therapeutic purposes. Myxin (a copper complex known commercially as Cuprimycin) is produced from iodinin and has been authorized for the treatment of superficial bacterial infections for veterinary uses (Turner and Messenger, 1986).

The antagonistic effect of phenazine compounds is by no means limited to microbial cultures. Inhibition of uptake of molecular oxygen by mouse monocytes, HeLa cells, isolated mouse liver mitochondria and pig peritoneal macrophages, has been reported in the literature. More recently an interesting role has been suggested for phenazines in the cytostatic effect of *Ps. aeruginosa* and its supernatants in lymphocyte cultures. Such an effect may be related to the local immunosuppression mechanisms at sites of chronic infection by this bacterium (Sorensen and Klinger, 1987).

Phenazine compounds have recently received some interest from the pharmacological viewpoint. A hypotensive activity has been ascribed to some phenazine molecules (especially 3,6-dihydroxyphenazine-1-carboxylic acid). It was shown that the conversion of angiotensin I to angiotensin II by the angiotensin converting enzyme (ACE) may be inhibited by this phenazine compound. Angiotensin II is an active pressor substance which may be the causative agent in several forms of hypertension (high blood pressure).

A patent has been granted to Kingston *et al.* (1986) regarding the production of such a phenazine ACE inhibitor by a fermentation process with a strain of the *Streptomyces tanashiensis-zaomyceticus* group.

7.1 BIOSYNTHESIS

The search for a phenazine precursor to supply the aromatic moiety of phenazines led to the proposal of anthranilic acid.

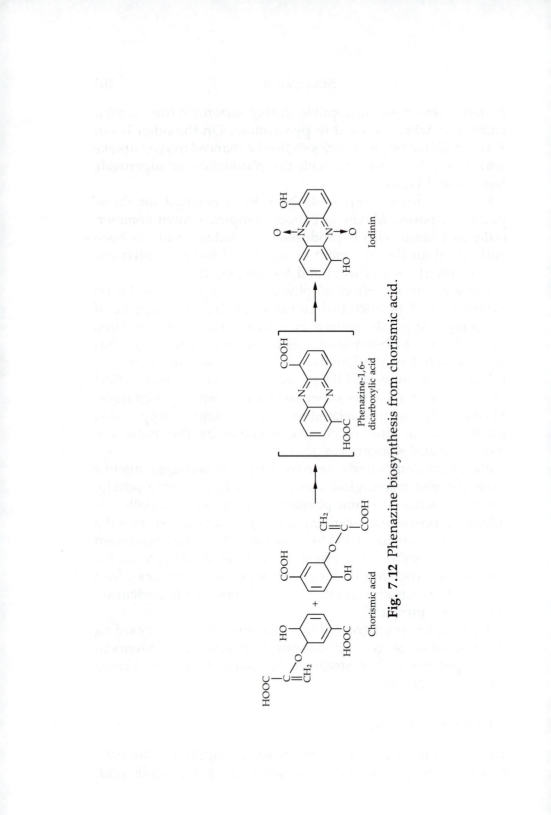

Fig. 7.12 Phenazine biosynthesis from chorismic acid.

Chorismic acid

Phenazine-1,6-dicarboxylic acid

Iodinin

However, this could not be verified. Instead shikimic acid, a common precursor of aromatic compounds was identified. Employing mutants of *Ps. aeruginosa* unable to degrade shikimic acid, Ingledew and Campbell (1969) showed that 98% of the phenazine formed was provided by shikimic acid. Similar results were reported for phenazine-1-carboxylic acid (*Ps. aureofaciens*) and iodinin (*Brevibacterium iodinum*). Further work with mutants blocking different stages in the branched pathway to aromatic amino acids (tryptophan and anthranilic acid) suggested that the branchpoint for phenazine biosynthesis was chorismate (Longley *et al.*, 1972).

Much work has been done on the problem of how chorismic acid is assembled to form the phenazine skeleton. Most workers seem to agree that two molecules of chorismic acid form the intermediate phenazine-1,6-dicarboxylic acid which is the key molecule for the phenazine family of compounds (Herbert *et al.*, 1976; Etherington *et al.*, 1979). The nitrogen of the phenazine molecule is probably derived from the amide of glutamine as the primary nitrogen source for the ring assembly. So far no nitrogen-containing intermediate after chorismate, and prior to the formation of phenazine-1,6-dicarboxylic acid, has been agreed upon. Herbert *et al.* (1982) suggested the intermediate shown in Fig. 7.13, phenazine-1,6-dicarboxylic acid, is the key intermediate, the metabolism to the type specific compound(s) seems to be variable with different bacteria and Actinomycetes.

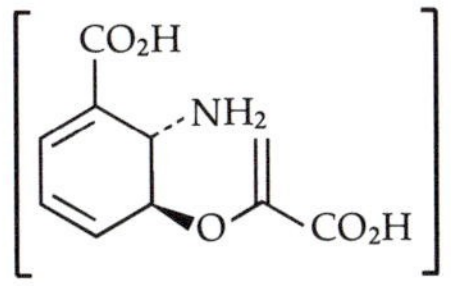

Fig. 7.13 Hypothetical intermediate.

7.2 PHYSIOLOGICAL SIGNIFICANCE

Studies on the kinetics of phenazine production by various bacteria has unequivocally revealed the 'secondary metabolite' nature of these compounds. This was further corroborated by evidence from work with different media that showed that catabolite repression of phenazine biosynthesis takes place in the presence of certain carbon and nitrogen sources. Derepression of phenazine biosynthesis by certain amino acids which are not incorporated into the phenazine molecule further emphasizes the regulatory mechanism involved.

Why do these micro-organisms produce phenazine so abundantly? Do they serve any particular purpose in the survival of these organisms? Being secondary metabolites, we may attempt to answer this question by pointing out that many microbial metabolites are not necessarily produced under natural conditions and may be, at least to a certain extent, the result of a strained metabolism coping with an unfavourable environment.

Additional information on the physiology of the phenazine compounds may be found in the review by Leisinger and Margraff (1979).

8

Other heterocyclic pigments

8.1 PRODIGIOSIN

The red pigment of *Serratia marcescens* (*Bacterium prodigiosum*) is well known to all bacteriologists. This organism, once believed to be part of the pigment-producing Chromobacteriaceae, is now classified with the Enterobacteriaceae. The genus is characterized by its attractive brick red pigment. The description of the 'polenta bleeding' organism by Bizio at the beginning of the 19th century was probably the first attempt to explain a phenomenon known during earlier days as 'blood in food' and the inevitable search for the malefactors concerned.

Prodigiosin is apparently also produced by some bacteria from marine sources (*S. marinorubra* and the psychrophil *Vibrio psychroerythreus*). Among the Actinomycetales *Nocardia madura*, *N. pelletierie*, *Streptomyces longisporubei* and *Streptoverticillium* were described as producers of prodigiosin or prodigiosin-like pigments (Gerber and Lechevalier, 1976).

It was the pyrrole structure of prodigiosin that attracted the attention of many physiologists and chemists. Although pyrroles constitute many of the primary, life essential compounds in animals and plants (haem, bile pigments, chlorophylls, phytochrome) their occurrence in microbial systems is quite unusual, as no physiological function can be ascribed to them. Furthermore, the tripyrrole structure of prodigiosin is quite unique (Fig. 8.1).

Prodigiosin is a secondary metabolite with a considerably narrow temperature range for production (24–28 °C). When

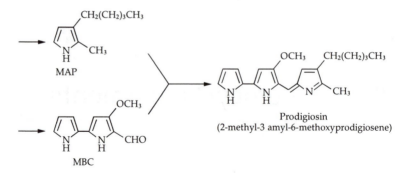

Fig. 8.1 Precursors for prodigiosin biosynthesis (after Williams, 1973).
MAP = 2-methyl-3-amylopyrrol,
MBC = 4-methoxy-2,2'-bipyrrole-5-carboxaldehyde.

prodigiosin-producing bacteria are grown on casein hydrolysate media the idiophase takes place between 48–96 h. Some nutrient factors such as thiamin and ferric ions are needed for optimal pigment production. An anionic detergent was found to stimulate prodigiosin formation. An increase in the exposure of negative binding sites for the fixation of the basic pyrrole pigment or precursor has been suggested for this enhancement (Feng *et al.*, 1982).

S. *marcescens* has been known as a saprobic bacterium. However during the last few years hospital infections due to S. *marcescens* have been recorded. It is not known whether pigment production may be involved in these infections. Since non-pigmented strains show a similar growth pattern it was assumed that the prodigiosin pigment plays no physiological role in this organism (Williams, 1973). It would be interesting to know whether pigmentless strains may also be involved in hospital infections.

The biogenesis of prodigiosin has been studied employing a number of strains which were complementary with respect to the formation of the pigment. The final step in the biosynthesis of prodigiosin is probably the condensation of 4-methoxy-2,2-bipyrrole-5-carboxaldehyde (MBC) with 2-methyl-3-amylpyrrol (MAP). This could be concluded from the fact that some strains produced MAP only, while others formed MBC only.

Synthrophic growth would lead to the formation of prodigiosin. From labelling experiments it was learned that proline
and glycine constitute most of the A and B pyrroles while ring
C pyrrole is mostly derived from acetate (Tanaka *et al.*, 1972).
The tripyrrole pigment is deposited in the cell envelope and is
not released into the medium, but may be extracted with
organic solvents. In its amorphous state it has the appearance
of red platelets with a green metallic sheen. It has a max. at
525–537 nm and shows some antibiotic activity against Gram
positive bacteria, some fungi as well as *Endamoeba histolytica*.

8.2 VIOLACEIN

Another nitrogenous heterocyclic pigment has been described
in *Chromobacterium violaceum*. It is a dark blue to violet pigment
that is formed in the bacterial cell, from which it may be
extracted by organic solvents. A lactose-containing medium
and prolonged incubation at 22 °C were found to be suitable for
pigment production.

Violacein is made of 3 subunits: 5-hydroxyindole, oxindole
and 2-pyrrolidone moieties (Fig. 8.2). A pigment having a
dimeric structure of two indole rings is not a frequent event in
nature and is therefore quite unusual in the microbial world.
The pyrrolidone nucleus is a derivative of the pyrrole ring and
thus shows some relationship to the prodigiosin molecule.
Having two indole nuclei it was assumed that tryptophan
would serve as the precursor. This was verified by work with
labelled substrates which showed that the carbon, nitrogen and
hydrogen atoms were derived exclusively from L-tryptophan,
while all the oxygen originated from molecular oxygen

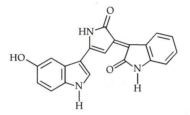

Fig. 8.2 Violacein.

(Hoshino *et al.*, 1987). The biosynthesis of violacein thus requires a rearrangement of the indole ring, decarboxylation and oxygenation, while the α-amino nitrogen of tryptophan serves for the formation of the pyrrolidone ring. The sequence of these reactions in the formation of the violacein pigment has not yet been worked out.

Antibiotic activity of violacein has been described against some Gram positive bacteria. However, as in the case of prodigiosin, no medical uses have been recorded.

8.3 BACTERIAL INDIGO

The brilliant blue pigment originally extracted from plants (leaves of *Indigofera tinctoria*) was widely used in ancient times for the dyeing of cotton and wool fabrics. Nowadays the chemically synthesized indigo is employed as a dye. It is interesting to note that plant indigo, as used in ancient times, is an artefact. The plant extract contains a colourless glucoside, indican, which, upon hydrolysis and oxidation, becomes the intense indigo colour (Fig. 8.3). Indigo is absorbed by the fabric only in the leuco (reduced) form. It is most likely that this reduction was carried out by microbial activity in the production vats, prior to the oxidative, colour yielding step, after exposure to air.

Indigo can also be produced by bacteria. As early as 1928, Gray described the formation of a pigment by *Pseudomonas indoloxidans*. This reaction was subsequently shown to involve the oxidation of indole to indoxyl by molecular oxygen. The indole substrate was derived from tryptophan supplied in the growth medium. This pigment was called indigotine, but was

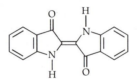

Fig. 8.3 Indigo.

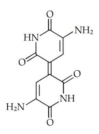

Fig. 8.4 Bacterial indigo (indigoidine).

claimed to be identical to the plant pigment. Indigotine was also described in the mould *Schizophyllum commune* (Miles *et al.*, 1956).

Another group of dark blue pigments of the indigo family was shown to be formed by various taxonomically unrelated bacteria. Bacterial indigo or indigoidine may be obtained from cultures of *Corynebacterium insidosum, P indigofera, Arthrobacter atrocyaneus* and *A. polychromogenus*, as well as the plant pathogen *Erwinia chrysanthei* (Knackmuss, 1973). Although these pigments were described during the early days of bacteriology, there was an uncertainty with regard to the chemical nature of indigoidine and its relationship to indigo. This was eventually cleared up by Elazari-Volcani (1939) who convincingly demonstrated the chemical differences between indigotine and indigoidine (Fig. 8.4). Bacterial indigo is formed extracellularly and is insoluble in many organic solvents, but may be dissolved in various concentrated acids with a change in colour (indigo blue in concentrated HCl; yellow in concentrated HNO_3). The diazadiphenoquinone structure was suggested by Kuhn *et al.* (1965).

Another famous dye, the Tyrian purple, was prepared in ancient times from the purple snail (*Murex brandaris*). This pigment is a bromo derivative of indigo, resulting from the oxidation by air of the colourless fluid of the snail glands. It is not known whether, in this case, microbial activity was also involved.

While most bacterial indigo is produced by colonies grown on plates containing conventional media (potato–glucose–peptone agar) in some cases it was observed that pigment

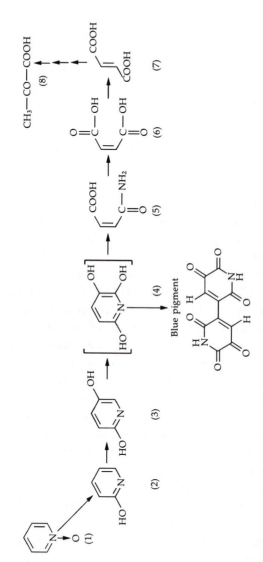

Fig. 8.5 Suggested pathway for the formation of a blue pigment by *Nocardia sp.* when cultivated in media containing pyridine-*N*-oxide (PNO) or 2-hydroxypyridine (adapted from Shukla and Kaul, 1986).

(1) Pyridine-*N*-oxide,
(2) 2-hydroxypyridine,
(3) 2,5-dihydroxypyridine,
(4) 2,3,6-trihydroxypyridine,
(5) maleamic acid,
(6) maleic acid,
(7) fumaric acid,
(8) pyruvic acid.

production requires the presence of 2-hydroxypyridine (*A. crystallopoietes*). In this case a dark green pigment is formed, the product of the acid hydrolysis of indigoidine. Little is known about the biosynthetic pathways that lead to the formation of these pigments. However, more recently (Shukla and Kaul, 1986) an attempt has been made to study the metabolism of 2-hydroxypyridine as well as other pyridines by a *Nocardia sp.* isolated from soil. A crystalline blue-black pigment was produced when this organism was grown on phosphate agar containing 0.2% of pyridine-*N*-oxide or 2-hydroxypyridine. The pigment was identified as 4.5,4′.5′-tetrahydroxy-3.3′-diazadiphenoquinone (2,2′), reminiscent of bacterial indigo. The suggested biosynthetic pathway is shown in Fig. 8.5. This of course does not explain the fact that some organisms produce

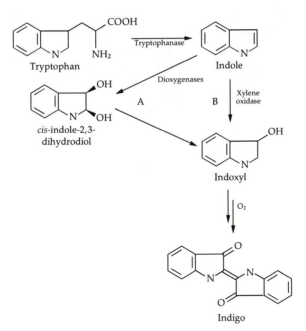

Fig. 8.6 Indigo biosynthesis. (A) Dioxygenase route to indoxyl by *E. coli* with cloned genes encoding a naphthalene dioxygenase enzyme from *Pseudomonas putida*. (B) Hydroxylase route to indoxyl catalysed by a *Pseudomonas TOL* plasmid-encoded xylene oxidase (adapted from Mermod *et al.*, 1986).

these pigments on simple media while others require the incorporation of pyridines.

There has been renewed interest in the biotechnological production of indigo (indigotine) employing genetically engineered bacteria. Indigo was shown to be formed by *E. coli* cells containing cloned genes encoding a naphthalene dioxygenase enzyme of *Pseudomonas putida*. The pathway involves the formation of indole from tryptophan (*P. indoloxidans*) followed by the formation of *cis*-indole-2,3-dihydrodiol from indole by a naphthalene dioxygenase. Spontaneous dehydration of the indole-2,3-dihydrodiol results in the formation of indoxyl, which, in turn, oxidizes spontaneously to form indigo. When a *xyl*A gene is inserted (TOL plasmid) indole is directly oxidized to indoxyl (xylene oxidase), which is transformed to indigo (Mermod *et al.*, 1986).

A hundred years ago, indigo was an expensive dye shipped from distant lands where indigo plants were grown. Since the beginning of this century indigo dyes have been prepared synthetically on a large scale and have been employed extensively in the textile industry. Will the new biotechnological process substitute the chemical synthesis of indigo?

9

Microbial naphthoquinones

Quinones display an array of colours: yellow, orange or red according to the position of the keto groups. Literature abounds on quinoid pigments isolated from various plant sources as well as microbial systems. They may be classified in different groups according to their chemical structure. The benzoquinones, such as the dark purple pigment spinulosin (Fig. 9.1) in certain species of Penicillium and Aspergillus, the naphthoquinones, for example javanicin (Fig. 9.2) in Fusarium species, and the anthraquinones, such as emodin (Fig. 9.3) from *P. islandicum*. They may be of an even more complex nature, as the red pigment piloquinone (Fig. 9.4) of *Streptomyces pilosus*. Most of these pigments and their biosynthetic

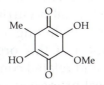

Fig. 9.1 Spinulosin.

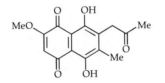

Fig. 9.2 Javanicin.

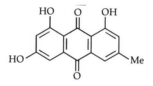

Fig. 9.3 Emodin.

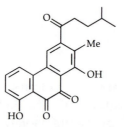

Fig. 9.4 Piloquinone.

pathways have been reviewed by Thomson (1976). Here I shall concentrate on the naphthoquinones which have been thoroughly studied more recently.

In fungi, the naphthoquinone pigments are very common and have been the subject, not only of research into their chemical structure, but also into their biosynthesis and biological significance. Fusarium species have been the organisms of choice in these studies. Over 30 different naphthoquinone compounds have been described so far, all occurring in the widespread *F. solani* and related fungi, including *Nectria haematococca*, the ascomycetous perfect form of this organism. Fusarubin (Fig. 9.5), javanicin (Fig. 9.2) and related compounds have displayed a number of physiological properties comprising fungitoxicity,

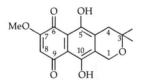

Fig. 9.5 Fusarubin.

antibacterial activities, insecticidity and phytotoxicity, as well as membrane modification and metal chelating characteristics (Parisot *et al.*, 1990).

Biosynthetic studies have shown that naphthoquinones like javanicin are derived from acetate (polyketide) and methionine, the latter yielding the methoxy group of the molecule. However the sequence of reactions leading to various naphthoquinones is still unclear. Various compounds have been described during the early period of growth, giving rise eventually to javanicin after only 7 days. However, it is not certain that this is due to a true metabolic sequence.

A number of media have been examined and their ability to support the growth of the fungus and the production of the naphthoquinones has been studied. A glucose–mineral salts medium with ammonium sulphate as a nitrogen source, supplemented with zinc and magnesium ions was found to be optimal for pigmentation. Naphthoquinone biogenesis is apparently favoured by a drop in pH to around 3 when ammonium sulphate is employed. Similar pH drops with ammonium phosphate were not accompanied by pigment formation, which may be due to repression by excessive phosphate availability. Naphthoquinone biosynthesis may be promoted by small amounts of aspartic or glutamic acid (Kern, 1978) as well as 5-fluorouracil (5-FU) when added to a glucose–asparagine medium for the cultivation of *N. haematococca*. It was suggested that 5-FU may interfere with nucleic acid metabolism and thus lead to naphthoquinone overproduction.

Mutation work with *F. solani* and *N. haematococca* yielded a number of overproducing strains, mostly red pigmented mycelia, which cause a dark red colouration of the medium due to the release of water soluble naphthoquinone pigments into the environment. Little has been published so far on the biotechnological parameters that control a naphthoquinone fermentation, although the genes that control pigment formation have already been described (Parisot *et al.*, 1984).

The contribution of naphthoquinone pigments to the pathogenicity of *F. solani* to plants has been studied by a number of workers with a view of finding some biological role

for these pigments. Direct application of Fusarium naphthoquinones indeed led to damage in the plasma membrane of the
plant cells with leakage of amino acids, nucleotide, proteins
and minerals from the cell. Chloroplast membranes were dis-
rupted and chlorophylls were released (Kern, 1978). However
this cannot be considered proof of the involvement of
naphthoquinones in the pathogenicity of *F. solani* to plants.
Marcinkovska *et al.* (1982) were unable to demonstrate a correlation between naphthoquinone content and pathogenicity.
Neither could genetic analysis indicate such a correlation. By
crossing various strains of *N. haematococca*, Holenstein and
Defago (1983) obtained a recombinant strain showing high
pathogenicity to pea plants but little or no naphthoquinone
production. Being typical secondary metabolites, the biological
role of naphthoquinones in fungi remains obscure.

At this juncture, I should like to recall the well-known observation of mycologists and most mushroom collectors that many
Boletus species develop a bluish colouration when the mushroom is damaged. This phenomenon evaded a biochemical
explanation for many years until Steglich and his group (1968)
showed that this change of colour was due to a pigment of the
vulpinic acid group, the yellow variegatic acid (Fig. 9.6), which
contains a catechol system in ring A. Exposure to air stimulates
an enzymic reaction which leads to the formation of the blue
quinone methide anion (Fig. 9.6).

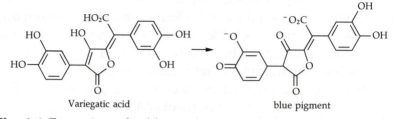

Variegatic acid   blue pigment

Fig. 9.6 Formation of a blue quinone methide ion from variegatic
acid.

10

Monascus pigments

Water soluble pigments from microbial sources are not so common but are of great biotechnological interest. The current antagonism towards synthetic colours has invited a great deal of research into the potential of micro-organisms to provide suitable water soluble pigments for use as colourants in the food industry. Water soluble yellow, orange and red pigments are currently of great interest.

In the Far East, coloured products such as red rice wine (Shao-Hsing) and native foods, such as red soy bean cheese, have been known for thousands of years. Ang-Kak or red rice is another well known item in the Chinese cuisine. In all these cases a fungal organism of the Monascus type is involved.

Monascus is a typical Ascomycete that produces a cleisto-thecium, a closed fruiting body containing eight ascospores, but reproduces asexually by the formation of conidiospores and a vegetative mycelium. Many species and strains have been described and appear under various names in most culture collections. The best known is *M. purpureus*. This and other species (*M. anka*, *M. ruber* and *M. paxii*) were found to produce a number of pigments of similar chemical structure. The elemental composition of some of the Monascus pigments are described in Table 10.1. Note the difference with regard to the nitrogen content. Some of the suggested structures for a number of Monascus pigments are given in Fig. 10.1.

M. purpureus produces such pigments on solid media or foodstuffs (e.g. moistened rice on flat trays). Reports on the

Table 10.1 Elemental composition of some Monascus pigments

Name	Empirical formula	Colour
Ankaflavin	$C_{23}H_{30}O_5$	Yellow
Monascin	$C_{21}H_{26}O_5$	Yellow
Monascorubin	$C_{23}H_{26}O_5$	Orange
Rubropunctamin	$C_{21}H_{22}O_5$	Orange
Monascorubramin	$C_{23}H_{27}NO_4$	Red
Rubropunctamin	$C_{21}H_{23}NO_4$	Red

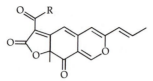

Fig. 10.1 Rubropunctatin: $R = n - C_5H_{11}$; Monascorubrin: $R = n - C_7H_{15}$.

production of these pigments by submerged fermentation have also been published (Lin, 1973; Yoshimura *et al.*, 1975; Carels and Shepherd, 1977). When grown on suitable media containing a carbon source (glucose) an organic nitrogen source (peptone or amino acids), at pH 6–7 and 25–28 °C, the mycelium grows in the form of small pellets which yields good pigmentation within 130–150 h. However, submerged fermentations yield only small amounts in comparison to solid-state culture. Evans and Wang (1984) suggested that support of the mycelium may explain the enhanced pigment production when the fungus is grown on a solid substrate.

The pigments may appear both in the mycelium and the fermentation broth. Extraction of the mycelium with methanol yields two dominant colours: a yellow component (peak near 390 nm) and a red pigment (peak near 500 nm). Most of the pigments can be found in the mycelium where they are formed; only a fraction is solubilized in the medium. A number of workers have pointed out that the amount of pigment found in the aqueous broth was dependent on the amount of water soluble proteins or amino acids available in the solution (Broder and Koehler, 1980).

The type of pigments formed by the mycelium may be in response to environmental conditions, which also affect quantitatively the biosynthetic reactions. At a 15% glucose level most of the pigment was of the red type. Ground rice was effective in producing the red pigment, while corn and potato starch were more conducive to orange and red-orange colouration.

Monascus pigments may gain importance in the near future. Attempts to reduce or substitute the nitrate in the curing of meat because of the hazardous nitrosamine, may lead to the colouration of meats by such fungal pigments. So far little has been published on the commercial production of Monascus pigments or their use as food colourants. A monosodium glutamate complex of Monascus pigments showed considerable colour stability in response to drastic pH changes. However, thermal stability is apparently not so good. Autoclaving water soluble pigments resulted in a red cloudy solution, indicating a precipitation of the protein–pigment complex.

11

Microbes and non-microbial pigments

The pigment-forming potential of many microbes is not always apparent, since a suitable substrate is not always available. Microbes may be involved in reactions that yield products that affect other substrates, yielding coloured material. The most important example of this kind is the process of curing, the production of colour in meat.

Salting and curing was originally employed to preserve meat at times when no other means (chilling, freezing) of preservation were available. Nowadays, the cured colour and aroma have become of greater importance. Saltpetre (nitrate) was the major vehicle for preservation and curing. Not until the beginning of the present century was it realized that, before curing can take place, nitrite must be produced from nitrate. Hence, nitrite is the actual curing agent. The conversion of nitrate (NO_3) to nitrite (NO_2) is a well known microbial process, nitrate serving as an electron acceptor during the anaerobic respiration of a variety of micro-organisms.

In an acid environment, nitrite yields nitric oxide (NO) which reacts with the muscle pigment (myoglobin) or blood pigment (haemoglobin). The red cured colour thus formed (nitrosomyoglobin or nitrosohaemoglobin) is relatively more stable to light and oxygen, but very stable to heat. Thus, cooked cured sausages conserve their characteristic pink colour, in contrast to uncured meat that becomes greyish-brown (Fig. 11.1). Since

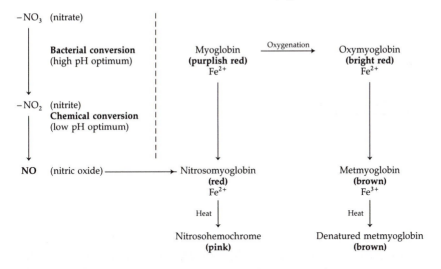

Fig. 11.1 Chemical changes in cured and non-cured meats.

NO_2 is the actual curing agent, this can be added directly to the curing mixture. For health security reasons this must be done in the form of nitrite curing salt (about 0.5% $NaNO_2$ in NaCl). For further information on the technology of meat curing the interested reader is referred to a recent review by Wirth (1986).

What are the micro-organisms involved in the reduction of nitrate to nitrite? Many aerobic bacteria under microaerophilic conditions will use nitrate as an electron acceptor for respiration. During the initial phases of curing, high concentrations of Micrococci (10 × other bacteria) have been found. These Micrococci are Gram positive, catalase positive, salt and nitrite tolerant bacteria that appear in clusters or packets of coccoidal cells and have an oxidative metabolism in contrast to the similar, but more fermentative Lactobacilli, which increase in number with time. Both groups of bacteria are probably also involved in the formation of the characteristic cured flavour of such meats.

12

Epilogue

It was suggested earlier that microbial pigments having no obvious function should be classified as secondary metabolites. However, we may ask again, why are these compounds produced at all? Can we dismiss this question simply by assuming a regulatory mechanism that went astray, yielding structurally complex products that are of no use to the producer? Or should we better claim, with some modesty, that secondary metabolites are compounds produced by certain micro-organisms, with physiological functions still unclear, but which may be elucidated by further studies?

A typical case which may illustrate this point is *Pseudomonas fluorescens*. Every bacteriologist has come across this organism that abundantly populates water and soil. Under certain growth conditions, a water soluble, yellow-green fluorescent pigment (pyoverdin, in analogy to pyocyanine, the phenazine pigment of *P. aeruginosa*) is produced. Several chemical structures have been suggested but none has been agreed upon. Neither has there been any role ascribed to the pyoverdin pigment. However, pyoverdin can be produced in very large amounts (over 1 mg/g dry weight). Can this be just futile biosynthesis?

Meyer and Abdallah (1978) studied the formation of pyoverdin in various media. When grown in a synthetic medium with succinate as the sole carbon source, large amounts of the pigment were produced. When 1 mg/l Fe^{3+} ions were added there was a doubling of the growth yield, with a complete repression of pyoverdin formation. But when Fe^{3+} was growth limiting (less than 200 μg Fe^{3+}/l) there was an inverse relationship between the iron content and the amount of pyoverdin synthesized after entry into the stationary phase of growth. Indeed, succinate grown cells (without the addition of ferric ions) soon became iron deficient and produced large amounts of pyoverdin. With citric or malic acids as substrate, no pigment was formed, since no iron starvation could be achieved unless a specific iron chelator was added.

The specific derepression of pyoverdin synthesis which results from iron limitation suggests that the pigment does play a specific role, either in the transport or the metabolism of iron. This is further supported by the fact that the fluorescent pigment is a powerful chelator of Fe^{3+} ions (not Fe^{2+}) with a high affinity constant. Pyoverdin contains the unusual amino acid σ-N-hydroxyornithine in a cyclic peptide chain. Similar structures have also been found in other siderophores which are known to facilitate iron transport into microbes. Pyoverdin is certainly not a secondary metabolite.

References

Achenbach, H., Kohl, W. and Reichenbach H. (1974) Structure of flexirubin. *Tetrahedron Lett.*, **30**, 2555.

Andrews, A.G., Phaff, H.J. and Starr, M.P. (1976) Carotenoids of *Phaffia rhodozyma*, a red pigmented fermenting yeast. *Phytochemistry*, **15**, 1009–11.

Aoshima, I., Tozawa, Y., Ohmono, S. and Ueda, K. (1985) Production of decolorizing activity for molasses pigment by *Coriolus versicolor* Ps4a. *Agric. Biol. Chem.*, **49**, 2041–5.

Arai, T. and Mikami, Y. (1972) Chromogenicity of Streptomyces. *Appl. Microbiol.*, **23**, 402–6.

Arcamone, F., Camerino, B. and Cotta, E. (1969) New carotenoids from *Streptomyces mediolani. n.sp. Experientia*, **25**, 241–2.

Aronson, J.N. and Vickers, R.S. (1965) Conversion of *m*-tyrosine to dihydroxyphenylalanine by a tyrosine hydroxylase from *Bacillus cereus. Biochim. Biophys. Acta*, **110**, 624–6.

Arpin, N. (1968) Les carotenoids des Discomycetes: Essai chimiotaxinomique. Thèse de doctorat, Univ. de Lyon.

Ashikawa, I., Miyata, A., Koike, H. and Koyama, Y. (1986) Light induced structural change of β-carotene in thylakoid membranes. *Biochemistry*, **25**, 6154–60.

Barnett, T.A. and Hageman, J.H. (1983) Characterization of a brown pigment from *Bacillus subtilis* cultures. *Can. J. Microbiol.*, **29**, 309–15.

Baron, S.S. and Rowe, J.J. (1981) Antibiotic action of pyocyanin. *Antimicrobial Agents Chemother.*, **20**, 814–20.

Battersby, A.R. (1988) Biosynthesis of the pigments of life. *J. Nat. Products*, **51**, 629–42.

Bauman, R. and Kocher, H.P. (1974) Genetics of *Streptomyces glaucescens* and regulation of melanin production. Second International Symposium on Genetics of Industrial Microorganisms. (ed. K.D. MacDonald), Academic Press, London, 535–51.

Bell A.A. and Wheeler, M.H. (1986) Biosynthesis and functions of fungal melanins. *Ann. Rev. Phytopathol.*, **24**, 411–51.

Bell, A.A., Puhalla, J.E., Tolmsoff, W.J. and Stipanovic, R.D. (1976) Use of mutants to establish scytalone as an intermediate in melanin biosynthesis by *Verticillium dahliae*. *Can. J. Microbiol.*, **22**, 787–99.

Ben-Amotz, A. and Avron, M. (1983) On factors which determine massive β-carotene accumulation in the halotolerant alga *Dunaliella bardawil*. *Plant Physiol.*, **86**, 1286–91.

Ben-Amotz, A., Mokay, S., Edelstein, S. and Avron, M. (1989) Bioavailability of natural isomer mixture as compared with all-*trans* β-carotene in rats and chicks. *J. Nutrition*, **119**, 1013–9.

Bendix, S. and Allen, M.B. (1962) Ultra-violet induced mutants of *Clorella pyrenoidosa*. *Arch. Mikrobiol.*, **41**, 115–41.

Bergman, K., Burke, P.V., Cerdá-Olmedo *et al.* (1969) *Phycomyces*. *Bacteriol. Rev.*, **33**, 99–157.

Blakeslee, A.F. (1904) Sexual reproduction in the Mucorinae. *Proc. Am. Acad. Arts Sci.*, **40**, 205–319.

Blois, M.S. (1978) The melanins: their synthesis and structure. *Photochem. Photobiol. Rev.*, **3**, 115–34.

Bloomfield, B.J. and Alexander, M. (1967) Melanins and resistance of fungi to lysis. *J. Bacteriol.*, **93**, 1276–80.

Blum, H.F. (1941) *Photodynamic Action and Diseases Caused by Light*. Van Nostrand Reinhold, New York.

Boekelheide, K., Graham, D.G., Mize, P.D. and Jeffs, P.W. (1980) The metabolic pathway catalyzed by the tyrosinase of *Agaricus bisporus*. *J. Biol. Chem.*, **255**, 4766–71.

Braeunlich, K. (1978) The chemistry and action of pigments in poultry diets. *World Poultry Congr.*, **15**, 236–40.

Bramley, P.M. and Mackenzie, A. (1988) Regulation of carotenoid biosynthesis. *Curr. Top. Cell. Regul.*, **29**, 291–345.

Britton, G. (1983) *The Biochemistry of Natural Pigments*. Cambridge University Press, Cambridge.

Broder, C.U. and Koehler, P.E. (1980) Pigments produced by *Monascus purpureus* with regard to quality and quantity. *J. Food Sci.*, **45**, 561–9.

Brockman, H. and Eggers, H. (1958) Partial synthesis of proto-hypericin from penicilliopsin. *Chem. Ber.*, **91**, 81–100.

Bu'Lock, J.D., Jones, B.E. and Winskill, N. (1976) The apocarotenoid system of sex hormones and prohormones in Mucorales. *Pure Appl. Chem.*, **47**, 191–202.

Burchard, R.P. and Dworkin, M. (1966) Light induced lysis and carotenogenosis in *Myxococcus xanthus*. *J. Bacteriol.*, **91**, 535–45.

Burgeff, H. (1924) Untersuchungen über Sexualität und Parasitismus bei Mucorineen. *Bot. Abhandl.*, **4**, 5–155.

Butler, M.J., Lazarovits, G., Higgins, V.J. and Lachance, M.A. (1989) Identification of a black yeast isolated from oak as belonging to the genus *Phaeococcomyces sp.* Analysis of melanin produced by the yeast. *Can. J. Microbiol.*, **35**, 728–34.

Canto de Loura, I., Dubacq, J.P. and Thomas, J.C. (1987) The effect of nitrogen deficiency on pigments and lipids of cyanobacteria. *Plant Physiol.*, **83**, 838–43.

Carels, M. and Shepherd, D. (1977) The effect of different nitrogen sources on pigment production and sporulation of Monascus species in submerged shaken culture. *Can. J. Microbiol.*, **23**, 1360–72.

Carra, P.O. and hEocha, C.Ó. (1976) Algal biliproteins and Phyco-bilins, in *Chemistry and Biochemistry of Plant Pigments*, 2nd edn (ed. T.W. Goodwin), Academic Press, London, 328–76.

Cerdá-Olmedo, E. (1985) Carotene mutants of *Phycomyces*. *Methods in Enzymol.*, **110**, 220–43.

Ciegler, A. (1965) Microbial carotenogenesis. *Adv. Appl. Microbiol.*, **7**, 1–34.

Claes, H. (1960) Interaction between chlorophyll and carotenes with different chromophoric groups. *Biochem. Biophys. Res. Commun.*, **3**, 585–90.

Claes, H. and Nakayama, T.C.M. (1959) The photooxidative fading of chlorophyll *in vitro* in the presence of carotenes with various chromophores. *Zeitschr. Naturforsch.*, **14B**, 746–7.

Clayton, R.K. (1970) *Light and Living Matter. Vol 1: The Physical Part*, McGraw-Hill, New York, 5–61.

Coggins, C.W., Henning, G.L. and Yokoyama, H. (1970) Lycopene accumulation induced by 2-(4-chlorophenylthio)-triethylamine hydrochloride. *Science*, **169**, 1589–90.

Crameri, R., Ettlinger, L., Hutter, R. *et al.* (1982) Secretion of tyrosinase in *Streptomyces glaucescens*. *J. Gen. Microbiol.*, **128**, 371–9.

Crippa, R., Horak, V., Prota, G. *et al.* (1989) Chemistry of melanins. *The Alkaloids*, **36**, 253–323.

Cripps, C., Bumpus, J.A. and Aust, S.D. (1990) Biodegradation of azo and heterocyclic dyes by *Phanaerochaeta chrysosporium*. *Appl. Environ. Microbiol.*, **56**, 1114–18.

Cubo, M.T., Buenida-Claveria, A.M., Beringer, J.E. and Ruiz-Sainz, J.E. (1988) Melanin production by rhizobium strains. *Appl. Environ. Microbiol.*, **54**, 1812–17.

Davies, B.H. (1973) Carotene biosynthesis in fungi. *Pure Appl. Chem.*, **35**, 1–28.

Della Cioppa, G., Garger, S.J., Sverlow, G.G. *et al.* (1990) Melanin production in *Escherichia coli* from a cloned tyrosinase gene. *Biotechnology.*, **8**, 634–8.

Demain, A. (1972) Riboflavin oversynthesis. *Annu. Rev. Microbiol.*, **26**, 369–88.

Di Mascio, P., Kaiser, S. and Sies, H. (1989). Lycopene as the most efficient biological carotenoid singlet oxygen quencher. *Arch. Biochem. Biophys.*, **274**, 532–8.

Dundas, I.D. and Larsen, H. (1962) The physiological role of the carotenoid pigments of *Halobacterium salinarium*. *Arch. Mikrobiol.*, **44**, 233–9.

Dworkin, M. (1958) Endogenous photosensitization in a carotenoidless mutant of *Rhodopseudomona spheroides*. *J. Gen. Physiol.*, **41**, 1099–112.

Elazari-Volcani, B. (1939) On *Pseudomonas indigofera* (Voges) Migula and its pigment. *Arch. Mikrobiol.*, **10**, 343–58.

Ellis, D.H. and Griffith, D.A. (1974) The location and analysis of melanins in the cell walls of some soil fungi. *Can. J. Microbiol.*, **20**, 1379–86.

Erikson, K.E. and Kirk, T.K. (1985) Biopulping biobleaching and treatment of Kraft bleaching effluents with white-rot fungi, in *Comprehensive Biotechnology* (ed. M. Moon-Young), Vol. 4, Pergamon Press, New York, 271–94.

Etherington, T., Herbert, R.B., Holliman, F.G. and Sheridan, J.B. (1979) The biosynthesis of phenazines: Incorporation of (^{2}H) shikimic acid. *J. Chem. Soc., Perkins Trans.*, **1**, 2416–19.

Evans, P.J. and Wang, H.Y. (1984) Pigment production from immobilized *Monascus sp.* utilizing polymeric resin adsorption. *Appl. Environ. Microbiol.*, **47**, 1323–6.

Feng, J.S., Webb, J.W. and Tsang, J.C. (1982) Enhancement by dodecylsulfate of pigment formation in *Serratia marcescens*. *Appl. Environ. Microbiol.*, **43**, 850–3.

Fiasson, J.L., Petersen, R.H., Bouchez, M.P. and Arpin, N. (1970)

Contribution biochimique à la connaissance taxinomique de certains champignons cantharelloides et clavariodes. *Rev. Mycol.*, **34**, 357–64.

Geis, P.A., Wheeler, M.H. and Staniszlo, P.J. (1984) Pentaketide metabolites of melanin synthesis in the dematiaceous fungus *Wangiella dermatitidis*. *Arch. Microbiol.*, **137**, 324–8.

Gerber, N. and Lechevalier, M.P. (1976) Prodiginine (prodigiosin-like) pigments from Streptomyces and other aerobic Actinomycetes. *Can. J. Microbiol.*, **22**, 658–67.

Giese, A.C. (1981) The photobiology of Blepharisma. *Photochem. Photobiol. Rev.*, **6**, 139–80.

Gill, M., Gimenez, A. and McKenzie, R.W. (1988) Pigments of fungi, Part 8. Biantheraquinones from *Dermocybe austroveneta*. *J. Nat. Products*, **51**, 1251–4.

Glazer, A.N. (1982) Phycobilisomes: Structure and dynamics. *Ann. Rev. Microbiol.*, **36**, 173–98.

Goedheer, J.C. (1965) Fluorescence decline in purple bacteria resulting from carotenoid absorption. *Biochim. Biophys. Acta*, **94**, 606–9.

Goldstrohm, D.D. and Lily, V.G. (1965) The effect of light on the survival of pigmented and nonpigmented cells of *Dacryopinax spathularia*. *Mycologia*, **57**, 612–23.

Gooday, G.W., Fawcett, P., Green, D. and Shaw, G. (1973) The formation of fungal sporopollenin in the zygospore wall of *Mucor mucedo*: A role for the sexual carotenogenesis in the Mucorales. *J. Gen. Microbiol.*, **74**, 233–9.

Goodchild, N.T., Kwock, L. and Lin, P.S. (1981) Melanin: a possible cellular superoxide scavenger, in *Oxygen and Oxy-radicals in Chemistry and Biology* (eds M.A.J. Rodgers and E.L. Powers), Academic Press, London, 645–8.

Goodwin, T.W. (1952) Carotenogenesis III. Identification of the minor polyene components of the fungus *Phycomyces blakesleeanus* and a study under various cultural conditions. *Biochem. J.*, **50**, 550–8.

Goodwin, T.W. (ed.) (1976) *Chemistry and Biochemistry of Plant Pigments*, Academic Press, London.

Goodwin, T.W. (1980) *The Biochemistry of the Carotenoids. Vol. 1. Plants*, 2nd edn, Chapman and Hall, London.

Goodwin, T.W. and Jamikorn, M. (1956) Studies in carotenogenesis XVIII. Carotenoid production by a strongly chromogenic bacterium isolated from butter. *Biochem. J.* **62**, 275–81.

Gray, P.H.H. (1928) The formation of indigotin from indol by soil bacteria. *Proc. R. Soc. Lond.*, **B102**, 2263–79.

Gregory, K.F. and Huang, J.C.C. (1964) Tyrosinase inheritance in *Streptomyces scabies*. *J. Bacteriol.*, **87**, 1284–6.

Griffiths, M., Sistrom, W.R., Cohen-Bazire, G. and Stanier, R.Y. (1955) Function of carotenoids in photosynthesis. *Nature*, **176**, 1211–14.

Haller, H.D. and Finn, R.K. (1979) Biodegradation of 3-chlorobenzoate and formation of black color in the presence and the absence of benzoate. *Eur. J. Appl. Microbiol.*, **8**, 191–205.

Halpern, Y.S., Teneh, R.B. and Grossowicz, F.G. (1962) Further evidence for the production of pyocyanine by nonproliferating suspensions of *Pseudomonas aeruginosa*. *J. Bacteriol.*, **83**, 935–6.

Hanson, A.W. (1968) Crystal structure of myxin. *Acta Crystallogr.*, **B24**, 1084–96.

Herbert, R.B., Holliman, F.G. and Sheridan, J.B. (1976) Biosynthesis of microbial phenazines: Incorporation of shikimic acid. *Tetrahedron Lett.*, 639–42.

Herbert, R.B., Mann, J. and Römer, A. (1982) Phenazine and phenoxazinone biosynthesis in *Brevibacterium iodinum*. *Zeitschr. Naturf.* 'C', **37**, 159–64.

Holenstein, J.E. and Defago, G. (1983) Inheritance of naphthazarin production and pathogenicity to pea in *Nectria haematococca*. *J. Exp. Bot.*, **34**, 297–35.

Hoshino, T., Takano, T., Hori, S. and Ogasawara, N. (1987) Biosynthesis of violacein: Origins of hydrogen, nitrogen and oxygen atoms in the 2-pyrrolidone nucleus. *Agric. Biol. Chem.*, **51**, 2733–41.

Howard, R.J. and Ferrari, M.A. (1989) Role of melanin in appressorium function. *Exp. Mycol.*, **13**, 403–18.

Humbeck, K. (1990) Photoisomerization of lycopene during carotenogenesis in mutant C–6D of *Scenedesmus obliquus*. *Planta*, **182**, 204–10.

Hynes, M.F., Brucksch, K. and Priefer, U. (1988) Melanin production encoded by a cryptic plasmid in a *Rhizobium leguminosarum* strain. *Arch. Mikrobiol.*, **150**, 326–32.

Ichiyama, S., Shimokata, K. and Tsukamura, M. (1988) Relationship between mycobacterial species and their carotenoid pigments. *Microbiol. Immunol.*, **32**, 473–9.

Ingledew, W.M. and Campbell, J.J.R. (1969) Evaluation of shikimic acid as a precursor of pyocyanine. *Can. J. Microbiol.*, **15**, 535–41.

Isler, O. (ed.) (1971) *The Carotenoids*. Birkhauser Verlag, Basel.

Ito, S. (1986) Reexamination of the structure of eumelanin. *Biochim. Biophys. Acta*, **883**, 155–61.

Ito, Y., Nanba, H. and Kuroda, H. (1979) Melanin produced by *Cochliobolus miyabenanus*. *Yakugaku Zasshi*, **99**, 1027–30.

Jayaram, M., Presti, D. and Delbrück, M. (1979) Light induced carotene synthesis in Phycomyces. *Exp. Mycol.*, **3**, 42–52.

Jenkins, C.L., Andrews, A.G., McQuade, T.J. and Starr, M.P. (1979) The pigment of *Pseudomonas paucibilis* is a carotenoid (nostoxanthin), rather than a brominated aryl-polyene (Xanthomonadin). *Curr. Microbiol.*, **3**, 1–4.

Johnson, E.A., Lewis, M.J. and Grau, C.R. (1980) Pigmentation of egg yolks with astaxanthin from the yeast *Phaffia rhodozyma. Poultry Sci.*, **59**, 1777–82.

Jones,O.T.G. (1978) Biosynthesis of porphyrins, hemes and chlorophylls, in *The Photosynthetic Bacteria* (eds R.K. Clayton and W.R. Sistrom), Plenum Press, New York.

Katz, E., Thompson, C.J. and Hopwood, D.A. (1983) Cloning and expression of the tyrosinase gene from *Streptomyces antibioticus* in *Streptomyces lividans. J. Gen. Microbiol.*, **129**, 2703–14.

Kern, H. (1978) Les naphthazarines des Fusarium. *Ann. Phytopath.*, **10**, 327–45.

Kingston, K.B., Slusarchyk, D.S., Mead, B. and Liu, W.C. (1986) Phenazine ace inhibitor produced from Streptomyces species, US Patent 4 318 987.

Knackmuss, H.J. (1973) Chemistry and biochemistry of azaquinones. *Ang. Chem. (Inter.)*, **85**, 163–9.

Kohl, W., Achenbach, H. and Reichenbach, H. (1983) Investigations on metabolites of microorganisms: Part 24. The pigments of *Brevibacterium linens*: aromatic carotenoids. *Phytochemistry*, **22**, 207–10.

Krinsky, N.I. (1964) Carotenoid de-epoxidations in algae: 1. Photochemical transformation of antheraxanthin to zeaxanthin. *Biochem. Biophys. Acta*, **88**, 487, 491.

Krinsky, N.I. (1968) The protective function of carotenoid pigments, in *Photophysiology*, Vol.III, (ed. A. Giese), Academic Press, London, 123–95.

Krinsky, N.I. (1989), Carotenoids as chemopreventive agents. *Prev. Med.*, **18**, 592–602.

Kubo, Y. and Furusawa, I. (1986) Localization of melanin in appressoria of *Colletotrichum lagenarium. Can. J. Microbiol.*, **32**, 280–2.

Kubo, Y., Furusawa, I. and Shishiyama, J. (1987) Relation between pigment intensity and penetrating ability in appressoria of *Colletotrichum lagenarium. Can. J. Microbiol.*, **33**, 870–3.

Kuhn, R., Starr, M.P., Kuhn, D.A. *et al.* (1965) Indigoidine and other bacterial pigments related to 3,3'-bipyridyl. *Arch. Mikrobiol.*, **51**, 71–84.

Kunisawa, R. and Stanier, R.Y. (1958) Studies on the role of carotenoid pigments in a chemoheterotrophic bacterium, *Corynebacterium poinsettiae*. *Arch. Microbiol.*, **31**, 146–56.

Kuo, M.J. and Alexander, M. (1967) Inhibition of the lysis of fungi by melanins. *J. Bacteriol.*, **94**, 624–9.

Kurtz, M.B. and Champe, S.P. (1982) Purification and characterization of the conidial laccase of *Aspergillus nidulans*. *J. Bacteriol.*, **94**, 1338–45.

Kwong-Chung, K.J. and Rhodes, J.C. (1986) Encapsulation and melanin formation as indicators of virulence in *Cryptococcus neoformans*. *Infect. Immun.*, **51**, 218–23.

Leisinger, T. and Margraff, R. (1979) Secondary metabolites of the fluorescent Pseudomonads. *Microbiol. Rev.*, **43**, 422–42.

Lerch, K. and Ettlinger, L. (1972) Purification and characterization of a tyrosinase from *Streptomyces glaucescens*. *Eur. J. Biochem.*, **31**, 427–37.

Liaaen-Jensen, S. (1965) Biosynthesis and function of carotenoid pigments in microorganisms. *Annu. Rev. Microbiol.*, **19**, 163–81.

Liaaen-Jensen, S. (1979) Carotenoids: A chemosystematic approach. *Pure Appl. Chem.*, **51**, 661–75.

Lin, C.F. (1973) Isolation and cultural conditions of *Monascus sp.* for the production of pigment in a submerged culture. *J. Ferm. Technol.*, **51**, 407–14.

Lipson, E.D. (1980) Sensory transduction in Phycomyces photo-responses, in *The Blue Light Syndrome* (ed. H. Senger), Springer Verlag, Berlin, 110–18.

Longley, R.P., Halliwell, J.E., Campbell, J.J.R and Ingledew, W.M. (1972) The branchpoint of pyocyanine biosynthesis. *Can. J. Microbiol.*, **18**, 1357–63.

Lopez-Diaz, I. and Cerdá-Olmedo, E. (1980) Relationship of photo-carotenogenesis to other behavioural and regulatory responses in *Phycomyces*. *Planta*, **150**, 134–9.

Macmillan, J.D., Maxwell, W.A. and Chichester, C.O. (1966) Lethal photosensitization of microorganisms with light from a continuous wave laser. *Photochem. Photobiol.*, **5**, 555–65.

Mann, S. (1969) Uber Melaninbildende Stämme von *Pseudomonas aeruginosa*. *Arch. Mikrobiol.*, **65**, 359–79.

Marcinkovska, J., Kraft, J.M. and Marquis, L. (1982) Phototoxic effect of cell-free cultural filtrates of *Fusarium solani* isolates on virulence, host specificity and resistance. *Can. J. Plant Sci.*, **62**, 1027–35.

Mason, H.S. (1948) The chemistry of melanin. *J. Biol. Chem.*, **172**, 83–99.

Mathews, M.M. (1963) Comparative study of lethal photosensitization of *Sarcina lutea* by 8-methoxysporalen and by toluidine blue. *J. Bacteriol.*, **85**, 322–8.

Mathews, M.M. (1964) The effect of low temperature on the protection by carotenoids against photosensitization in *Sarcina lutea*. *Photochem. Photobiol.*, **3**, 75–7.

Mathews, M.M. and Sistrom, W.R. (1960) The function of the carotenoid pigments of *Sarcina lutea*. *Arch. Mikrobiol.*, **35**, 139–48.

Maxwell, W.A., Macmillan, J.D. and Chichester, C.O. (1966) Function of carotenoids in protection of *Rhodotorula glutinis* against irradiation from a gas laser. *Photochem. Photobiol.*, **5**, 576–77.

Mayer, A.M. (1987) Polyphenol oxidases in plants – recent progress. *Phytochemistry*, **26**, 11–20.

Mencher, J.R. and Heim, A.H. (1962) Melanin biosynthesis by *Streptomyces lavendulae*. *J. Gen. Microbiol.*, **28**, 655–70.

Mermod, N., Harayama, S. and Timmis, K.N. (1986) New route to bacterial production of indigo. *Biotechnology*, **4**, 321–3.

Meyer, J.M. and Abdallah, M.A. (1978) The fluorescent pigment of *Pseudomonas fluorescens*: Biosynthesis, purification and physico-chemical properties. *J. Gen. Microbiol.*, **107**, 319–28.

Miles, P.G., Lund, H. and Raper, J.R. (1956) Identification of indigo as a pigment produced by a mutant culture of *Schizophyllum commune*. *Arch. Biochem. Biophys.*, **62**, 1–5.

Moore, M.M., Breedveld, M.W. and Autor, A.P. (1989) The role of carotenoids in preventing oxidative damage in the pigmented yeast, *Rhodotorula mucilaginosa*. *Arch. Biochem. Biophys.*, **270**, 413–19.

Morgan, J.G., McGettigan, S. and Donlon, J. (1985) Tyrosin 2-hydroxylation by *Aeromonas salmonicida*. *Biochem. Soc. Trans.*, **13**, 464–5.

Müller, R. and Schicht, G. (1965) Zur photosensibilität der Bakterien. *Arch. Mikrobiol.*, **51**, 290–306.

Murillo-Aranjo, F.J., Calderon, I.L. and Cerdá-Olmedo, E. (1982) β-carotene production strains of the fungus *Phycomyces blakesleeanus*. US Patent 4 318 987.

Nicolaus, R.A. (1968) Melanins, in *Chemistry of Natural Products* (ed. S. Lendered), Harman, Paris.

Ninet, L., Renaut, J. and Tissier, R. (1969) Activation of the biosynthesis of carotenoids by *Blakeslea trispora*. *Biotechnol. Bioeng.*, **11**, 1195–210.

Ootaki, T. (1973) A new method for heterokaryon formation in Phycomyces. *Mol. Gen. Genet.*, **121**, 49–56.

Ourisson, G., Rohmer, M. and Anton, R. (1979) From terpenes to sterols: Macroevolution and microevolution. *Recent Adv. Phytochem.*, **13**, 131–62.

Parisot, D., Devys, M. and Barbier, M. (1990) Naphthoquinone pigments related to fusarubin from the fungus *Fusarium solani* (Mart.). *Sacc. Microbios.*, **64**, 31–47.

Parisot, D., Maugin, M. and Gerlinger, C. (1984) Genes controlling pigmentation in *Nectria hematococca. J. Gen. Microbiol.*, **130**, 1543–55.

Parn, P. and Seviour, R.J. (1974) Pigments induced by organo-mercurial compounds in *Cephalosporium diospyros. J. Gen. Microbiol.*, **85**, 228–36.

Piatelli, M., Fattorusso, E., Nicolaus, R.A. and Magno, S. (1965) The structure of melanins and melanogenesis. V. Ustilagomelanin. *Tetrahedron*, **21**, 3229–36.

Pierson, B.K. and Castenholtz, R.W. (1978) Photosynthetic apparatus and cell membranes of the green bacteria, in *Photosynthetic Bacteria* (eds R.K. Clayton and W.R. Sistrom), Plenum Press, New York.

Polacheck, I. and Kwong-Chung, K.J. (1988) Melanogenesis in *Cryptococcus neoformans. J. Gen. Microbiol*, **134**, 1037–41.

Polak, A. (1989) Melanin as a virulence factor in pathogenic fungi. *Mycoses*, **33**, 215–24.

Pomerantz, S.H. and Murthy, V.V. (1974) Purification and properties of tyrosinases from *Vibrio tyrosinaticus. Arch. Biochem. Biophys.*, **160**, 73–82.

Prota, G. (1988) Progress in the chemistry of melanins and related metabolites. *Med. Res. Rev.*, **8**, 525–56.

Rao, S. and Modi, V.V. (1977) Carotenogenesis: Possible mechanism of action of trisporic acid in *Blakeslea trispora. Experientia*, **33**, 31–3.

Rau, W. (1976) Photoregulation of carotenoid biosynthesis in plants. *Pure and Appl. Chem.*, **47**, 237–43.

Reichenbach, H. (1986) The myxobacteria: common organisms with uncommon behaviour. *Microbiol. Sci.*, **3**, 268–74.

Reichenbach, H. and Kleinig, H. (1984) Pigments of Myxobacteria, in *Myxobacteria* (ed. E. Rosenberg). Springer Verlag, New York.

Reichenbach, H., Kohl, W., Bottger-Vetter, A. and Achenbach, H. (1980) Flexirubin type pigments in Flavobacterium. *Arch. Mikrobiol.*, **126**, 291–3.

140   References

Ross, I.K. (1979) *Biology of the Fungi*. McGraw-Hill, New York.

Rottem, S. and Markowitz, O. (1979) Carotenoids act as reinforcers of the *Acholeplasma laidlawii* lipid bilayer. *J. Bacteriol.*, **140**, 944–8.

Rowley, B.I. and Pirt, S.J. (1972) Melanin production by *Aspergillus nidulans* in batch and chemostat cultures. *J. Gen. Microbiol.*, **72**, 553–63.

Ruddat, M. and Garber, E.D. (1983) Biochemistry, physiology and genetics of carotenogenesis, in *Secondary Metabolism and Differentiation in Fungi* (eds J.W. Bennet and A. Ciegler), Marcel Dekker, New York.

Saiz-Jimenez, C. and Shafizadeh, F. (1985) Electron spin resonance spectrometry of fungal melanins. *Soil Sci.*, **139**, 319–25.

Sandmann, G., Bramley, P.M. and Boger, P. (1980) The inhibitory mode of action of the pyridazinone herbicide norflurazon on a cell-free carotenogenic enzyme system. *Pesticide Biochem. Physiol.*, **14**, 185–91.

Santamaria, L., Bianchi, A., Arnaboldi, A. *et al.* (1988) Chemoprevention of indirect and direct chemical carcinogenesis by carotenoids as oxygen radical quenchers. *Ann. N.Y. Acad. Sci.*, **534**, 584–96.

Schmidt, K. (1978) Biosynthesis of carotenoids, in *Photosynthetic Bacteria* (eds R.K. Clayton and W.R. Sistrom), Plenum Press, New York.

Schwartz, Y. and Margalith, P. (1965) Production of egg-yolk coloring material by a fermentation process. *Appl. Microbiol.*, **13**, 876–81.

Shiba, T. and Harashima, K. (1986) Aerobic photosynthetic bacteria. *Microbiol. Sci.*, **3**, 376–8.

Shirling, E. and Gottlieb, D. (1972) Cooperative description of type strains of *Streptomyces*. *Intern. J. System. Bacteriol.*, **22**, 265–394.

Shivprasad, S. and Page, W.J. (1989) Catechol formation and melanization by Na$^+$ dependent *Azotobacter chroococcum*: a protective mechanism for aeroadaptation? *Appl. Environ. Microbiol.*, **55**, 1811–17.

Shlomai, P., Ben-Amotz, A. and Margalith, P. (1991) Production of carotene stereoisomeres by *Phycomyces blakesleeanus*. *Appl. Microbiol. Biotechnol.*, **34**, 458–62.

Shropshire, W. (1980) Carotenoids as primary photoreceptors in blue light responses, in *The Blue Light Syndrome* (ed. H. Senger), Springer Verlag, Berlin.

Shukla, O.P. and Kaul, S.M. (1986) Microbial transformation of pyridine-N-oxide and pyridine by *Nocardia sp*. *Can. J. Microbiol.*, **32**, 330–41.

Sirianuntapiboon, S., Somchai, P., Ohmono, S. and Atthasampunna, P. (1988) Screening of filamentous fungi having the ability to decolorize molasses pigments. *Agric. Biol. Chem.*, **52**, 387–92.

Sisler, H.D. (1986) Control of fungal diseases by compounds acting as antipenetrants. *Crop Protection*, **5**, 306–13.

Smith, P.F. and Henrikson, C.V. (1966) Growth inhibition of Mycoplasma by inhibitors of polyterpene biosynthesis and its reversal by cholesterol. *J. Bacteriol.*, **9**, 1854–8.

Sorensen, R.U. and Klinger, J.D. (1987) Biological effects of *Pseudomonas aeruginosa* phenazine pigments. *Antibiot. Chemother.*, **39**, 113–24.

Stadnichuk, I.N., Romanova, N.I. and Seliakh, I.O. (1985) A phycourobilin containing phycoerythrin from the cyanobacterium *Oscillatoria sp. Arch. Microbiol.*, **143**, 20–5.

Stanier, R.Y. (1959) Formation and function of the photosynthetic pigment system in purple bacteria, in *The Photochemical Apparatus, its Structure and Function*, Brookhaven Symposium in Biology, **11**, Upton, New York, pp. 43–53.

Steglich, W., Furtner, W. and Prox, A. (1968) Neue Pulvinsäure-Derivate. *Zeitschr. Naturf.*, **23B**, 1044–50.

Stoeckenius, W. (1978) Bacteriorhodopsin, in *The Photosynthetic Bacteria* (eds R.K. Clayton and W.R. Sistrom), Plenum Press, New York.

Stormer, R.S. and Falkinham, J.O. (1989) Differences in antimicrobial susceptibility of pigmented and unpigmented colonial variants of *Mycobacterium avium. J. Clin. Microbiol.*, **27**, 2459–65.

Stüssi, H. and Rast, D.M. (1981) The biosynthesis and possible function of γ-glutaminyl-4-hydroxybenzene in *Agaricus bisporus. Phytochemistry*, **20**, 2347–52.

Sunada, K.V. and Stanier, R.Y. (1965) Observation on the pathway of carotenoid synthesis in Rhodopseudomonas. *Biochim. Biophys. Acta*, **107**, 38–43.

Sutter, R.P. (1970) Effect of light on β-carotene accumulation in *Blakeslea trispora. J. Gen. Microbiol.*, **64**, 215–21.

Sutter, R.P. (1975) Mutations affecting sexual development in *Phycomyces blakesleeanus. Proc. Nat. Acad. Sci. USA*, **72**, 127–30.

Sutter, R.P. and Whitaker, J.P. (1981a) Zygophore-stimulating precursors (pheromones) of trisporic acids active in *Phycomyces blakesleeanus. J. Biol. Chem.*, **256**, 2334–41.

Sutter, R.P. and Whitaker, J.P. (1981b) Sex pheromone metabolism in *Blakeslea trispora. Naturwiss.*, **68**, 147–8.

Suzue, G., Tsukada, K. and Tanaka, S. (1967) A new triterpenoid from a mutant of *Staphylococcus aureus*. *Biochim. Biophys. Acta.*, **144**, 186–8.

Takaichi, S., Shimada, K. and Ishidsu, J.I. (1990) Carotenoids from the aerobic photosynthetic bacterium, *Erythrobacter longus*: β-carotene and its hydroxyl derivatives. *Arch. Mikrobiol.*, **153**, 118–22.

Tanaka, W.K., Bascur de Medina, L. and Hearn, W.R. (1972) Labeling patterns in prodigiosin biosynthesis. *Biochem. Biophys. Res. Commun.*, **46**, 731–7.

Taylor, R.F. (1984) Bacterial triterpenoids. *Microbiol. Rev.*, **48**, 181–98.

Terao, J. (1989) Antioxidant activity of β-carotene related carotenoids in solution. *Lipids*, **24**, 659–61.

Thomson, R.H. (1976) Quinones: nature, distribution and biosynthesis, in *Chemistry and Biochemistry of Plant Pigments* (ed. T.H. Goodwin), Academic Press, London, 527–559.

Trias, J., Viñas, M., Guinea, J. and Lorèn, J.G. (1989) Brown pigmentation in *Serratia marcescens* cultures associated with tyrosine metabolism. *Can. J. Microbiol.*, **35**, 1037–42.

Turner, J.M. and Messenger A.J. (1986) Occurrence, biochemistry and physiology of phenazine pigments production. *Adv. Microbiol. Physiol.*, **27**, 211–75.

Valadon, L.R.G. (1976) Carotenoids as additional taxonomic characters in fungi: A review. *Trans. Br. Mycol. Soc.*, **67**, 1–15.

Van den Ende, H. and Stegwee, D. (1971) Physiology of sex in Mucorales. *Bot. Rev.*, **37**, 22–36.

Wanner, G., Formanek, H. and Theimer, R.R. (1981) The ontogeny of lipid bodies (spherosomes) in plant cells. *Planta*, **151**, 109–23.

Williams, R.P. (1973) Biosynthesis of prodigiosin, a secondary metabolite of *Serratia marcescens*. *Appl. Microbiol.*, **25**, 396–402.

Wirth, F. (1986) Curing: Color formation and color retention in frankfurter-type sausages. *Fleischwirtschaft*, **66**, 354–8.

Witz, D.F., Hessler, E.J. and Miller, T.L. (1971) Bioconversion of tyrosine into the propylhygric acid moiety of lincomycin. *Biochemistry*, **10**, 1128–33.

Wolkow, P.M., Sisler, M.D. and Vigil, E.L. (1983) Effects of inhibitors of melanin biosynthesis on structure and function of appressoria of *Colletotrichum lindemuthianum*. *Physiol. Plant Pathol.*, **22**, 55–72.

Woloshuk, C.P., Sisler, H.D. and Vigil, E.L. (1983) Action of the antipenetrant, tricyclazole, on appressoria of *Pyricularia oryzae*. *Physiol. Plant Pathol.*, **23**, 245–59.

Wright, L. and Rilling, H.C. (1963) The function of carotenoids in a photochromogenic bacterium. *Photochem. Photobiol.*, **2**, 339–42.

Yoneda, F. (1984) Vitamin B$_2$, in *Encyclopedia of Chemical Technology* (eds M. Grayson and D. Eckroth), 3rd edn, Vol. 24, (Kirk-Othmer), Wiley Interscience Publications, New York, 108–24.

Yoshimura, M., Yamanaka, S., Mitsugi, K. and Hirose, Y. (1975) Production of Monascus pigment in a submerged culture. *Agric. Biol. Chem.*, **39**, 1789–95.

Zhdanova, N.N., Gavryushina, A.I. and Vasilevskaya, A.I. (1973) Effect of γ and UV irradiation on survival of *Cladosporium sp.* and *Oidiodendron cerealis. Mikrobiol. Zh. (Kiev)*, **35**, 449–52.

Zink, P. and Fengel, D. (1988) Studies on coloring matter of blue stain fungi. *Holzforschung*, **42**, 217–20.

Author index

Species index

Numbers in **bold** refer to figures, numbers in *italics* refer to tables

Subject index

Numbers in **bold** refer to figures, numbers in *italics* refer to tables